# Fungal Biology

**Series Editors**

Vijai Kumar Gupta, Biorefining and Advanced Materials Research Center
Scotland's Rural College (SRUC), SRUC Barony Campus, Parkgate
Dumfries, Scotland, UK

Maria G. Tuohy, School of Natural Sciences
National University of Ireland Galway
Galway, Ireland

Fungal biology has an integral role to play in the development of the biotechnology and biomedical sectors. It has become a subject of increasing importance as new fungi and their associated biomolecules are identified. The interaction between fungi and their environment is central to many natural processes that occur in the biosphere. The hosts and habitats of these eukaryotic microorganisms are very diverse; fungi are present in every ecosystem on Earth. The fungal kingdom is equally diverse, consisting of seven different known phyla. Yet detailed knowledge is limited to relatively few species. The relationship between fungi and humans has been characterized by the juxtaposed viewpoints of fungi as infectious agents of much dread and their exploitation as highly versatile systems for a range of economically important biotechnological applications. Understanding the biology of different fungi in diverse ecosystems as well as their interactions with living and non-living is essential to underpin effective and innovative technological developments.This series will provide a detailed compendium of methods and information used to investigate different aspects of mycology, including fungal biology and biochemistry, genetics, phylogenetics, genomics, proteomics, molecular enzymology, and biotechnological applications in a manner that reflects the many recent developments of relevance to researchers and scientists investigating the Kingdom Fungi. Rapid screening techniques based on screening specific regions in the DNA of fungi have been used in species comparison and identification, and are now being extended across fungal phyla. The majorities of fungi are multicellular eukaryotic systems and therefore may be excellent model systems by which to answer fundamental biological questions.A greater understanding of the cell biology of these versatile eukaryotes will underpin efforts to engineer certain fungal species to provide novel cell factories for production of proteins for pharmaceutical applications. Renewed interest in all aspects of the biology and biotechnology of fungi may also enable the development of "one pot" microbial cell factories to meet consumer energy needs in the 21st century. To realize this potential and to truly understand the diversity and biology of these eukaryotes, continued development of scientific tools and techniques is essential. As a professional reference, this series will be very helpful to all people who work with fungi and should be useful both to academic institutions and research teams, as well as to teachers, and graduate and postgraduate students with its information on the continuous developments in fungal biology with the publication of each volume.

Alexandre Gomes Rodrigues • Arti Gupta
Editors

# Fungal Infections

## Causes, Diagnostics, and Treatments

Springer

*Editors*
Alexandre Gomes Rodrigues
NNIT Switzerland AG
Basel, Basel-Stadt, Switzerland

Arti Gupta
Smarak Degree College
Shri Avadh Raz Singh
Faizabad, Uttar Pradesh, India

ISSN 2198-7777 ISSN 2198-7785 (electronic)
Fungal Biology
ISBN 978-3-032-06013-6 ISBN 978-3-032-06014-3 (eBook)
https://doi.org/10.1007/978-3-032-06014-3



This Springer imprint is published by the registered company Springer Nature Switzerland AG
The registered company address is: Gewerbestrasse 11, 6330 Cham, Switzerland

# Contents

# Introduction to Fungal Infections

**Richa, Rishi Kumar Saxena, and Namrata**

## Introduction

### *Fungi*

In the Precambrian era, fungus first appeared as a unique class of single-celled eukaryotes. Fungi are classified as a diverse group of eukaryotic organisms that embraces yeasts, molds, and mushrooms. The English word "fungus" is precisely borrowed from the Latin word "fungus". Fungi are classified in a distinct kingdom, called Fungi, separate from bacteria, plants, and animals because of the presence of chitin in the cell wall as well as a distinct nucleus and organelles bound by a membrane. Molecular phylogeny strongly supports the idea that fungi belong to a single group of related creatures called Eumycota (also known as real fungi or Eumycetes), who share a common ancestor and form a monophyletic group (Money 2016).

### *Diversity in Fungi*

From unicellular aquatic chytrids to massive mushrooms, the fungus kingdom contains a vast range of taxa with distinct ecologies, life cycle methods, and morphologies. Though estimated at 2.2–3.8 million species, little is known about the true richness of the fungal world (Hawksworth and Lücking 2017). Out of these, only

Richa (✉) · R. K. Saxena
Department of Microbiology, Bundelkhand University, Jhansi, Uttar Pradesh, India

Namrata
Institute of Food science and Technology, Bundelkhand University,
Jhansi, Uttar Pradesh, India

 
A. Gomes Rodrigues, A. Gupta (eds.), *Fungal Infections*, Fungal Biology,
https://doi.org/10.1007/978-3-032-06014-3_1

roughly 148,000 have been described (Cheek et al. 2020), with at least 300 potentially harmful to humans and over 8000 known to be harmful to plants (Anonymous 2017). Fungi manifest themselves in a variety of morphological shapes and sizes, ranging from tiny yeasts to enormous, intricate mushrooms.

Additionally, certain fungi associate symbiotically with other living things. For example, lichens, are created when fungal and algae or cyanobacteria form a symbiotic association. Fungi exhibit a great deal of diversity in their morphology, which is indicative of their ability to adapt to a broad variety of ecological niches and lifestyles.

## Taxonomic Classification of Fungi

Generally, organisms are identified, described, and given names in taxonomic classification. They are then classified according to a system of groupings, such as genus, family, and order. Systemic biology is the study of the relationships between species and larger groups of organisms in terms of evolution (Money 2016).

The integration of DNA analysis into taxonomy has been made possible by developments in molecular genetics, which has occasionally called into question the historical classifications based on morphology and other characteristics. The fungus kingdom is classified into one subkingdom, phyla, and subphyla (Table 1). Phylogenetic studies that were published in the first decade of the twenty-first century have contributed to this reorganization of the taxonomic structure.

Fungi are categorized under the kingdom Fungi. Fungal taxonomy is determined by a number of factors, such as morphology, reproductive strategy, and genetic links. Fungi belong to the monophyletic group of opisthokonts, which is more closely linked to animals than to plants (Shalchian-Tabrizi et al. 2008). Molecular phylogenetic analyses verify that fungi have a single evolutionary ancestor (Hibbett et al. 2007; Li et al. 2016). Fungal taxonomy is always changing, particularly as a result of DNA comparison-based research. These modern phylogenetic investigations often overthrow the classifications created using older, perhaps less reliable methods based on physical traits and biological species theories developed from experimental matings (Anonymous 2024).

The kingdom Fungi contains additional, smaller phyla and groupings that demonstrate the diversity of this kingdom. As new information about the genetic linkages and ecological roles of fungi becomes available, the taxonomy and understanding of these organisms are constantly changing (Money 2016).

**Table 1** Kingdom Fungi categorized in different phylum with their major characteristics

| Phylum | Characteristics | Spores |
|---|---|---|
| Chytridiomycota | Chytrids are a class of fungi that are primarily aquatic and have spores that are flagellated, sometimes known as zoospores | Flagellate cell |
| Zygomycota | The creation of zygospores with suspensors that originate in merging of specific hyphae is a defining feature of zygomycetes | Zygospores |
| Ascomycota | Ascus or Ascospores are produced by ascomycetes, often known as sac fungus, inside structures called asci that resemble sacs. Yeasts, moulds, and numerous well-known fungus, such as morels and truffles, are included in this group | Ascus or Ascospores |
| Basidiomycota: | Club-shaped structures called basidia are the home of basidiospores, which are produced by basidiomycetes, also known as club fungi. Toadstools, bracket fungus, and mushrooms are members of this category | Basidiospores |
| Glomeromycota | Glomeromycetes play a crucial role in nutrient intake through their symbiotic associations like mycorrhizae with plant roots | Mycorrhizae |
| Microsporidia | Single-celled fungi known as microsporidians are frequently parasitic and infect a variety of animals, including people | Single-celled |
| Neocallimastigomycota | These fungi are specifically designed to decompose plant matter in herbivore digestive tracts, especially in ruminants such as cows | Uniflagellate zoospores |
| Blastocladiomycota | This category of fungi comprises both infectious and saprophytic species, distinguished by their biflagellate zoospores | Biflagellate zoospores |

## *Physiological Traits of Fungi*

Since the groundbreaking taxonomy studies of Carl Linnaeus, Christiaan Hendrik Persoon, and Elias Magnus Fries in the eighteenth and nineteenth centuries, fungi have been categorized based on their morphology (i.e., traits like spore colour or microscopic features) or physiology. Fungi can have very different morphologies based on the species and stage of life. Still, a number of general traits are shared by a wide variety of fungi. Fungus are composed of feathery filaments called hyphae. As they take up nutrients from food, they also release enzymes. Hyphae possess a hard chitin cell wall. Seeking other food sources, they develop from a tip and spread outward. Hyphae, structures resembling threads, comprise fungi. It is possible for these hyphae to be aseptate, or undivided and multinucleate, or septate, or divided into compartments by cross-walls made of septa.

Mycelia are huge networks made up of branching hyphae. A mycelium is a hyphae-rich substance that makes up a fungus's body. It is the primary feeding structure of the fungus and is in charge of taking up nutrients from the surroundings.

Spores are basically single cells or tiny clusters of cells with the ability to grow into new fungi given the correct circumstances. Depending on the species, fungus produce spores by either sexual or asexual means (Money 2016).

## Multiplicative or Reproductive Structures

A wide range of reproductive structures can be produced by fungi, such as fruiting tissue like mushrooms and sporangia, which are specialized structures with spore-producing cells inside.

## Microscopic Structures of Fungi

Most fungi, which are widely distributed, are undetectable due to their microscopic structures and secretive lives in soil or on decaying matter. Many fungi grow as hyphae, which are cylindrical, thread-like structures that can reach lengths of several centimetres and diameters of 2–10 μm. Hyphae emerge through their tips, or edges; new hyphae are usually created by new tips emerging along existing hyphae through a process known as branching, although sometimes growing hyphal tips fork, resulting in two hyphae that grow in parallel (Harris 2008).

A process known as hyphal fusion, also known as anastomosis, occurs occasionally when hyphae come into touch with one another. The consequence of these growth processes is the formation of a mycelium, which is a web of hyphae (Alexopoulos et al. 2024). Hyphae are classified as coenocytic or septate. Coenocytic hyphae lack cross-walls and remain continuous, while septate hyphae are divided into compartments by septa formed at right angles to the cell wall. Each compartment contains one or more nuclei. Septa include pores through which cytoplasm, organelles, and occasionally nuclei can flow. Dolipore septum found in fungi belonging to the phylum Basidiomycota is one example of septate fungi (Deacon 2005). Coenocytic hyphae can be thought of as multinucleate supercells (Chang and Miles 2004).

Several species have evolved specialized hyphal structures for absorbing of food from living hosts, such as arbuscules of various mycorrhizal fungi, which enter the host cells and absorb nutrients, and the next is haustoria, seen in plant-parasitic species of most fungal phyla (Bozkurt and Kamoun 2020; Parniske 2008).

## Macroscopic Structures

Fungal mycelia, often known as moulds, are visible to the unaided eye on a variety of substrates and surfaces, including wet walls and spoilt food. Colony is the term used to describe mycelia cultivated in the lab culture dishes on solid agar medium. Because of spores or pigmentation, these colonies may display different growth forms and colours or shades that can be utilized as diagnostics to distinguish between different species or groupings. Some single fungal colonies can grow to incredible sizes and ages. One such example is a clonal colony of *Armillaria solidipes* that is believed to be around 9000 years old and covers an area of more than 900 ha (Hanson 2008).

## *Reproduction in Fungi*

In general, the way the fungi grow and absorb nutrients sets them apart from all other living things, including animals. They grow from the tips of filaments (hyphae) that comprise the bodies of the creatures (mycelia) (Britannica 2024). Depending on the species and the surrounding circumstances, fungus can reproduce sexually or asexually.

### Asexual Reproduction

Asexual reproduction usually leads to the development of a genetic replica of the parent by one member without the genetic factor of another organism. The thallus, or fungal body, can be broken apart to produce new fungi, which is arguably the most basic mechanism of fungal reproduction. Asexual reproduction happens by mycelial fragmentation or vegetative spores termed "conidia". In comparison to sexual reproduction, mycelial fragmentation and vegetative spores enable more rapid dissemination while maintaining clonal populations suited to a particular habitat (Köhler et al. 2017). Species without apparent sexual cycle are referred to as "Fungi imperfecti" (fungi without the perfect or sexual stage), or Deuteromycota (Alcamo 2003).

Asexual reproduction also occurs by budding. In this process, certain single-celled fungi called yeasts reproduce by simple fission, where a single cell divides into two daughter cells after performing nuclear division; such cells then divide again once a certain development has occurred, and finally a population of cells forms. Fungi are typically grown in laboratories on a layer of solid nutritional agar that has been introduced with pieces or spores.

Spore formation is a typical asexual reproductive mechanism in fungus, though fragmentation, fission, and budding are also common. Large amounts of spores are usually produced and distributed by a variety of channels, including the water, the air, and animals. Fungi rely on asexual reproduction to survive and spread, as it enables them to quickly settle in new areas. Mitospores are a term used to describe asexually produced spores, which can be formed in a number of ways.

### Sexual Reproduction

The fungal reproductive system is distinct in numerous aspects. The fungal species is able to get used to new surroundings through sexual reproduction, which is a significant source of genetic variety. The process of sexual reproduction in fungi entails the union of specialized cells known as gametes, which vary in size and shape. A zygote is formed when gametes fuse together, and it goes through meiosis in order to produce spores. In fungi, sexual reproduction frequently entails intricate life cycles and can produce progeny with varying genetic makeup plasmogamy, karyogamy, and meiosis are the three successive steps of sexual reproduction in fungi (Britannica 2024).

It is not the same as sexual reproduction in plants or animals in many ways. Fungal groupings differ from one another as well, and these variations can be utilized to distinguish between different species based on physical variations in their reproductive and sexual structures (Guarro et al. 1999; White et al. 1990). Based on biological species principles, fungal isolates can identify species through mating tests (White et al. 1990). Earlier, the physical characteristics of the spores and sexual structures of the major fungal groups was used to distinguish them. For instance, the spore-containing structures, known as asci and basidia, can be used to identify ascomycetes and basidiomycetes, respectively. A dikaryotic stage is experienced in several ascomycetes and basidiomycetes, during which the nuclei inherited from the two parents stay separate in the hyphal cells following cell fusion. The complexity of fungal reproduction is a reflection of the genetic variations and lifestyle variations seen throughout this vast kingdom of living things (Alexopoulos et al. 2024).

Roughly one-third of all fungi are thought to reproduce by more than one means. For instance, a species' life cycle may have two distinct stages for reproduction i.e. the teleomorph, which is sexual reproduction, and the anamorph, which is asexual reproduction. Environmental signs set off genetically predetermined developmental phases that culminate in the formation of specialized reproductive systems, whether asexual or sexual. By effectively distributing spores or propagules carrying spores, these structures facilitate reproduction (Britannica 2024).

## *Fungi and Ecosystem*

Fungi play an essential and constructive role in the ecosystem. Together with recycling nutrients and decomposing organic materials as part of an ecosystem, they produce antimicrobial agents. However, certain fungi are detrimental, they significantly affect plant, human, and animal life. Infections caused by them can demolish human-made structures like buildings and mass-produced products. Similar to animals, fungi are heterotrophs. Several fungi act in symbiotic associations with plants to facilitate their uptake of nutrients and water from the soil. They have historically been used to ferment a variety of food products, including wine, beer, and soy sauce, as well as a direct source of food for humans in the form of truffles and mushrooms.

Fungi have been utilized to generate antibiotics since the 1940s, and more recently, other enzymes that they produce have found industrial and cleansing agent applications. Fungi are employed as biological insecticides to manage insect pests, plant diseases, and weeds. Alkaloids and polyketides are two examples of the bioactive substances known as mycotoxins, which are produced by many species and hazardous to all animals, including humans. Certain species' fruiting structures are used in conventional religious practices or for recreational purposes; they contain psychoactive chemicals. Crop losses brought on by fungi (such as rice blast disease) or food spoiling can significantly affect local economics and human food supplies.

Some most significant organisms in terms of their biological and socioeconomic contributions are found in the fungus kingdom. They maintain the flow of essential nutrients within ecological systems. Maximum vascular plants could not survive without the help of the symbiotic fungus like mycorrhizae, which reside in their roots and supply essential nutrients. Different fungi are responsible for producing a variety of drugs (including penicillin and other antibiotics), gourmet foods like mushrooms, truffles, and morels, as well as the bubbles in champagne, beer, and bread (Berkeley 2024).

## Fungal Infestation on Humans

Humans breathe in 1000–10 billion spores every day, and they are frequently confronted by enormous quantities of spores. Various fungal species live in multiple parts of the human body, such as the skin, gut, and other surfaces of the mucous membrane (Konopka et al. 2019).

Numerous harmful fungus are parasitic on humans and have the capacity to infect humans as well as other animals with diseases (Britannica 2024). Even while the direct harm that fungi offer to human health is concerning, the indirect threat that fungal diseases of plants provide to global food security is equally concerning. Fungi produce toxic substances, including those that cause terrible diseases, such as cancer, that negatively impact food sources and lower yields of regular crops (Konopka et al. 2019). There are up to 6 million species in the fungal kingdom (Taylor et al. 2014), and they have a profound and wide-ranging influence on biomedical research, manufacturing, agriculture, biodiversity, and global health.

Approximately 600 fungal species have been identified to be closely associated with humans, either as beneficial microbiota members or as infectious agents that may bring about some of the deadliest infectious diseases (Fisher et al. 2012, 2016, 2018).

## *Fungal Infections*

Human fungal infections have been on the rise in tandem with the introduction of novel, cutting-edge medical treatments such as immunosuppressive medications, antibiotics, and implanted medical devices (Konopka et al. 2019). Since human body temperature is a primary barrier to fungal diseases, despite the fact that numerous fungi are linked with people, relatively few of them are serious disease-causing agents (Robert and Casadevall 2009). The usual body temperature of humans and other mammals is high enough for most fungi to infiltrate, yet the fungal kingdom survives at ambient temperatures seen in nature.

Another reason fungal infections are such a serious problem is that they are very difficult to treat. Fungi are minute organisms that can dwell in the soil, water, air, plants, and even on human skin. They are the source of fungal infections. While the majority of fungi are benign, some can become infections when they invade the tissues and grow. Fungal infections can affect the skin, nails, hair, or internal organs including the bloodstream or lungs. Additionally, fungus cause a wide range of diseases that severely impact humans, they include athlete's foot, ringworm, and several other deadly conditions. Fungi and animals have numerous chemical and biological similarities than any other living creature, treating fungus-related diseases is extremely challenging. Many fungi serve as beneficial " exemplary microorganisms" for studying genetic and molecular biology-related topics, with yeasts being a notable exception (Berkeley 2024).

The majority of fungi are widespread and capable at multiplying without the assistance of humans or other animals in their natural habitats. Certain fungi might infect healthy individuals with sickness, but many species only turn harmful when the host is weak, such as when the immune system is weakened or compromised (Garber 2001). Numerous diseases in humans have been attributed to fungal infections, sometimes known as mycoses. Mycoses may vary in severity through superficial infections affecting the skin's surface layer to widespread infections affecting the kidneys, liver, spleen, brain, and lungs. Patients suffering from acquired immunodeficiency syndrome, individuals on cancer therapy, recipients of organ transplants, and patients undergoing major surgery are among the growing number of patients who are susceptible to invasive fungal infections (Walsh and Dixon 1996).

Invasive fungal infections are quite probable to occur in each of these patient groups. The range of opportunistic fungal pathogens that infect these patients keeps growing along with the population that is at risk. It can be challenging to identify severely invasive mycoses at an early stage and to treat them successfully. Extensive research is being conducted on the development of novel strategies for the detection and treatment of invasive fungal diseases (Walsh and Dixon 1996). It was fungi that infected the International Space Station (Satoh et al. 2016), moreover, black moulds are the only living organisms that have multiplied in the damaged ecosystems

**Table 2** Type of fungal infection and their causing agent

| Type of fungal infection | Disease | Causing agent | Origin/ region |
|---|---|---|---|
| Superficial mycoses | Dermatophytosis<br>Superficial candidosis | *Malassezia globosa*<br>*Pityriasis versicolor*<br>Others (e.g., Scopulariopsis, Scytalidium infections) | Tropical |
| Subcutaneous mycoses | Sporotrichosis<br>Mycetoma<br>Chromoblastomycosis<br>Phaeohyphomycotic cyst<br>Subcutaneous zygomycosis<br>Lobomycosis | *Sporothrix schenckii, C. albicans, Fonsecaea pedrosoi* | Tropical |
| Systemic mycoses | Systemic candidosis, aspergillosis, zygomycosis, fusariosis<br>Cryptococcosis | Opportunistic<br>Tropical specific forms (e.g., paranasal *Aspergillus granuloma*) | Tropical |
| Endemic | Histoplasmosis<br>African histoplasmosis<br>Blastomycosis<br>Coccidioidomycosis<br>Paracoccidioidomycosis | *Penicillium marneffei* | Tropical |

surrounding the Chernobyl nuclear power plant, which further demonstrates the flexibility of fungi to new situations (Casadevall et al. 2017).

## *Types of Fungal Infections*

Depending on the variety of fungus and the immune condition of the afflicted individual, there are numerous varieties of fungal infections observed. Several of these infections known as primary fungal pathogens, can cause illness in otherwise healthy, non-immune people as well. Invasive fungal infections pose a life-threatening risk in every situation (Brown et al. 2012a).

Fungal infections are majorly categorized into three groups (Table 2). Certain fungal species are pathogenic, producing infections in humans that may be superficial, subcutaneous, or systemic in nature. Most fungi that trigger deeply rooted systemic infections, enter the body primarily through breathing or lesions. Others, such as *Candida albicans*, are typical inhabitants of the skin and gastrointestinal system, but occasionally they can proliferate and enter the bloodstream after being put into the body by arterial stents or other medical devices (Garber 2001).

## Superficial Fungal Infections

It has been discovered that fungi can spread through mucous membranes, hair, nails, and the outermost layers of the skin. The incidence of these disorders has been gradually rising in recent years due to the rise in immunocompromised patients as well as the expansion of fitness facilities and public swimming areas, which encourage the spread of infection (Detandt and Nolard 1995; Garber 2001). Superficial infections are observed all over the world. There are geographical variations, especially when it comes to ringworm or dermatophyte infections that are widespread and pass from child to child. Several could have become more common and dispersed due to socio-economic or environmental factors in the tropics. *Pityriasis versicolor* is a common condition in tropical regions, mostly due to *Malassezia globosa*. Whereas these illnesses are also prevalent in environments with moderate temperatures (Hay 2006).

## Generally, Two Types of Superficial Fungal Infections Are Found

### Skin and Keratinized Tissue Infestations

An epidermal (skin) wound is the most common entry point for parasitic fungi into the human body. These wounds could be insect bites or unintentional scrapes, cuts, or bruises. *Claviceps purpurea*, the source of ergotism, or St. Anthony's fire, is one example of a fungus that can infect people.

During the Middle Ages, this disease was common in northern Europe, especially in areas where rye bread was consumed in large quantities. Ergot fungus spores are carried by the wind to the rye flowers, where they germinate, infect, and eradicate the plant's fruits, replacing them with masses of microscopic threads bonded together (Britannica 2024).

The lives of millions of individuals globally are impacted by superficial fungal infections, and they involve some of the most commonly reported skin infections (Piérard et al. 1996). The majority of superficial fungal infections are caused by dermatophytes, notably Trichophyton spp., Microsporum spp., and Epidermophyton spp., while yeasts and certain non-dermatophyte fungi can also be the culprits. Dermatophytes infect the stratum corneum of the epidermis as well as keratin-containing tissues like hair and nails (Table 3). The transmission of fungi is facilitated by getting into contact with infected people, objects, soil, or food. The position on the body of a particular dermatophyte disease determines the title it has (Garber 2001).

### Mucilageous Membrane Infections

The oesophagus, mouth, genitalia, or moist regions of the skin are all affected by mucosal candidiasis. In healthy women, vaginal candidiasis is the most common gynaecological infection, with *C. albicans* responsible for up to 95% of cases

**Table 3** Some common superficial fungal infections, observed in humans

| Disease | Causing agent | Infected body part | Diagnosis tools and management | References |
|---|---|---|---|---|
| *Tinea pedis* | *Trichophyton rubrum* or *T. mentagrophytes* var. *interdigitale*. | Nail infection | Patient health history, physical examination, and microscopy and culture of skin or nail specimens | Evans (1997), Semel and Goldin (1996) |
| *Tinea capitis* | *Microsporum audouinii* to *T. tonsurans* | Nail disorders or disease (In children over the age of 6 months) | Patient health history, physical examination, and microscopy and culture of skin or nail specimens | Bronson et al. (1983), Elewski (2000), Hay (2006), Richardson et al. (1995) |
| *Tinea versicolor* | *Malassezia furfur*, a commensal yeast | Infection of the skin (stratum corneum), nailzzs disorders | Patient health history, physical examination, and microscopy and culture of skin or nail specimens | Chiritescu et al. (1996), Garber (2001), Gupta et al. (1998), Levy (1997) |
| Onychomycosis | (i) *Dermatophytes*<br>(a) *T. rubrum*<br>(b) *T. mentagrophytes*<br>(ii) Non-dermatophyte yeasts (*C. albicans* and other non-*albicans* *Candida* spp.) and moulds (*Aspergillus* spp., *Fusarium* spp., *Acremonium* spp., *Scopulariopsis* spp. and *Scytalidium* spp.) | Toenail infections and 50% of fingernail infections (people with diabetes are at high risk) | Patient medical history, physical examination, and microscopy and culture of skin or nail specimens | Drake et al. (1998), Drake and Scher (1995), Ellis et al. (1997), Finlay (1997), Garber (2001), Greer (1995), Lubeck et al. (1993), Tosti et al. (2000) |

(Mendling et al. 1998; Regulez et al. 1994). Topical or oral azoles are ideal therapies for this easily diagnosable disease (Woolley and Higgins 1995).

More than 90% of people having advanced HIV infection will, at some point, acquire oropharyngeal candidiasis, a typical consequence of HIV infection that causes mucosal candidiasis (Vazquez 1999). Despite the fact that *Candida albicans* is the primary cause of this illness, several non-albicans Candida strains are now becoming more prevalent. Tactical and persistent azole-based medication is generally effective in treating oropharyngeal candidiasis (Powderly et al. 1999). However, in individuals with highly compromised immune systems who have recurrent HIV infection, treatment-resistant strains of Candida have been identified (Hoepelman and Dupont 1996). When a protease blocking medication is administered to HIV-positive patients, the incidence of the throat candidiasis is decreased (Diz-Dios et al. 1999).

Athlete's foot, ringworm, aspergillosis, histoplasmosis, and coccidioidomycosis are just a few numbers of human illnesses brought on by fungi. When in ecological equilibrium with other digestive system bacteria, the yeast *Candida albicans* a typical resident of the human mouth, throat, colon, and reproductive organs does not cause infection. But illness, ageing, and hormonal fluctuations can make *C. albicans* proliferate uncontrollably, overwhelming the body's defences and leading to candidiasis (also known as thrush when it affects the mouth) (Britannica 2024).

The manifestations of candidiasis can range from painful, inflammatory skin patches or elevated white patches on the tongue to a potentially fatal, invasive infection that affects the lining of the brain or heart. Human fungal illness incidence has increased as a result of better diagnosis and more worldwide travel, the latter of which has aided in the spread of tropical pathogenic fungus (Deorukhkar and Saini 2015).

## Subcutaneous Fungal Infections

Tropical regions and the subtropical regions are the primary locations for subcutaneous mycoses. Although such diseases are mostly limited to the skin and tissue beneath the skin, they can also spread to the outermost layer of skin and deeper tissues, such as bones. Typically, materials found in the outside world, including vegetation or soil, are used for insertion of microbes. Because of this point of entrance, these diseases have been referred to as the mycoses of the implantation process (Hay 2006).

Subcutaneous mycoses are mostly restricted to the dermis and subcutaneous tissues, though they might spread. Numerous pathogens, which are frequently limited to tropical and subtropical areas of the planet, are responsible for causing disease (Table 4). For instance, the dimorphic fungus *Sporothrix schenckii* is responsible for the development of sporotrichosis, which is the most common subcutaneous infection in some regions of Central and South Americas (Rios-Fabra et al. 1994). The fungus is present in soil and plants, and it typically infects farmers and gardeners,

**Table 4** Examples of subcutaneous fungal infection and their causing agent

| Subcutaneous fungal Disease | Causing agent | Human body part | References |
|---|---|---|---|
| Sporotrichosis | *Sporothrix schenckii* | Spread in dermis and subcutaneous tissues | Rios-Fabra et al. (1994) |
| Lymphocutaneous sporotrichosis | *Sporothrix schenckii* | Spread from lymphatic pathways to subcutaneous area of body | (Sharkey-Mathis 1993) |
| Chromoblastomycosis | *Fonsecaea pedrosoi*, *Cladosporium carrionii*, *F. compacta*, *Phialophora verrucosa* and *Rhinocladiella aquaspersa* | In advanced stage spreads to different organs like brain, lymph nodes, liver, lungs | Chapman and Daniel (1994) |
| Chronic mucocutaneous candidiasis | *C. albicans* | Found in skin and nails | Bonifaz et al. (1997) |

particularly people who look after roses, an infection appears as a lesion that can travel through the lymphatic pathways to other subcutaneous areas. Itraconazole treatment typically has a good effect on *lymphocutaneous sporotrichosis*, which is a non-life-threatening condition (Sharkey-Mathis 1993).

### Systemic Fungal Infections

Systemic fungal infections considered to be opportunistic generally arise once the host's defences are undermined in a certain way. These infections have a significant mortality rate and pose a threat to life. The incidence of systemic fungal infections is sharply rising due to the growing population of immunocompromised people (Denning et al. 1997; Groll et al. 1996). Furthermore, the human body's defences towards fungal pathogens are weakened by pharmacological therapy used to control the immune systems of cancer and transplant patients. Individuals afflicted with HIV, the virus that causes acquired immunodeficiency syndrome (AIDS), also have compromised immune systems against fungi, and fungal infections particularly those caused by *Aspergillus fumigatus* are a major cause of AIDS-related mortality.

Microbes responsible for systemic fungal infection can be classified into two groups: the real pathogenic (dimorphic) fungi, which are capable of penetrating and developing in the cells and tissues of a normal host without any discernible genetic predisposition, and the opportunists, such as *Aspergillus* and *Candida* spp. Opportunistic fungi are made up of more mild and less proficient organisms that can only penetrate the tissue of an immunodeficient patient, in contrast to dimorphic fungi. True pathogenic fungi, on the other hand, are limited in their geographic range. These fungi frequently cause moderate or asymptomatic infections that clear up quickly (Rautemaa-Richardson and Richardson 2017).

## Major Human Fungal Pathogens

Some common fungal human pathogens that have a large negative impact on the health of people and cause serious health issue are as follows.

## Candida

There are around 20 species of yeasts in the genus Candida that can infect people. In healthy individuals, these yeasts are typically found on mucous membranes, in the gut, and on the skin.

Candida has the potential to multiply and develop candidiasis when the body's defence system becomes compromised or when therapeutical agents eliminate bacteria that are normally present in the body (Kullberg and Arendrup 2015). It term to

**Table 5** Fungal infections caused by various *Candida* species in humans

| *Candida* infections | Candida sp. | Affected region/area | References |
|---|---|---|---|
| Candidiasis or candidemia | *C. albicans* | United States or North America spread in 95% area of region | Azie et al. (2012), Deorukhkar and Saini (2015) |
| | *C. glabrata* | | |
| | *C. parapsilosis* | | |
| | *C. krusei* | | |
| | *C. tropicalis* | | |
| | *C. kefyr* | | |
| | *C. dubliniensis* | | |
| | *C. famata* | | |
| | *C. rugose* | | |
| | *C. lipolytica* | | |
| | *C. norvegensis* | | |
| | *C. lusitaniae* | | |
| | *C. auris* | Spread in 30% area of region | Jeffery-Smith et al. (2018) |

this development as thrush as it typically occurs in the mouth, throat, or oesophagus. In individuals with AIDS brought on by HIV infection, thrush in the oesophagus is one of the most prevalent infections.

Vaginal yeast infections are another common symptom of Candida development, impacting approximately 75% of women at certain points in their lives. These infections are particularly common in women who are pregnant, have diabetes, are using antibiotics, or have had chemotherapy or medicines that depress their immune system. Therefore, both immunologically healthy and immunologically compromised people can become ill from *Candida* species (Table 5). Approximately 46,000 incidents of candidiasis are linked to healthcare facilities annually in the United States. Candida can also break free from the typical locations where it resides in our bodies and spread, resulting in a potentially fatal invasive disease (Tong et al. 2009). Candida infection, or candidemia as it is known, accounts for the majority of prevalent bloodstream infections in North America.

A nontoxic component of the human microbiome is *C. albicans*. But when the harmonious relationship between *Candida albicans* and host cells is upset, *C. albicans* develops on mucosal surfaces and/or invades cells of the host, it can spread into the bloodstream and colonize solid organs (Kim 2016).

## Aspergillus

Globally, the genus Aspergillus contains hundreds of species of mould that grow both inside and outside in nature. One and all are exposed for Aspergillus spores; these can be found in organic matter, moist environments, and cooling appliances. Aspergillus typically causes illness in people with weakened immune systems, lung injury, or severe allergies (Dagenais and Keller 2009). The term "aspergillosis" describes a group of illnesses brought on by Aspergillus, of which aggressive aspergillosis, persistent respiratory aspergillosis, aspergilloma, and allergic aspergillosis are the most prominent.

*Aspergillus* also serves as a common allergen that can lead to serious problems for those who have bronchiectasis, cystic fibrosis, and asthma. Although it can spread to other organs and tissues, invasive aspergillosis is most frequently seen in the lungs of those with severe immunosuppression. Between 25–90% of cases of invasive aspergillosis result in death (Pfaller and Diekema 2010). People with underlying lung diseases are typically the only ones affected by acute pulmonary aspergillosis, which is characterized by a persistent lung infection. It may show up as fungus-filled cavities in the lungs or even as an aspergilloma, which is a ball of fungus developing inside a cavity. Intense immunological reactions to *Aspergillus* infections in some people cause allergic aspergillosis, a persistent, crippling illness that can cause significant damage in those afflicted.

*A. fumigatus* is the most important fungus species linked to human illness, while other fungal species like *A. niger*, *A. flavus*, and *A. terreus* can also contribute to these illnesses. Since existing antifungal treatments for severe and prolonged aspergillosis conditions frequently fail to control these microbial infections, novel strategies for both treatment and prevention are required. Particularly, the development of triazole-resistant *A. fumigatus* has impeded the treatment progress reached thus far for severe aspergillosis. Apart from infecting humans, Aspergillus species may additionally infect animals, birds, and plants. They can also create chemicals that may lead food to deteriorate or trigger cancer.

Individuals breathe the airborne asexual spores of *A. fumigatus*, known as conidia, but in healthy persons, the innate immune system effectively eliminates them. Spores, despite this, can cause life-threatening aggressive aspergillosis in susceptible persons by eluding the immune system. The pathogenicity of *A. fumigatus* is not determined by a single trait; rather, the pathogen's capacity to withstand extreme temperatures, stress caused by oxidation, nutrient shortages, and hypoxia, along with its capacity to produce secondary metabolites and release enzymes for absorption of nutrients, all bring to the pathogen's survival within the body of an individual (Kim 2016).

## Cryptococcus

Patients with immunodeficiency are at risk of developing meningitis and fungal pneumonia due to the fungal human pathogen *C. neoformans*. There is proof that *C. neoformans* also infects well-immune hosts; however, healthy humans in tropical regions are infected by another similar species, *Cryptococcus gattii* (Panackal et al. 2015). Among human-related fungal infections, *C. neoformans* is distinct due to its polysaccharide encapsulation (Almeida et al. 2015; Bahn and Jung 2013; Choi et al. 2015; O'Meara and Alspaugh 2012).

Lungs or central nervous system is typically affected by cryptococcosis, which is typically brought on by two ways *Cryptococcus neoformans* or *Cryptococcus gattii*. Worldwide, *C. neoformans* is an expanding environmental microorganism that infects human beings through inhaling spores or dried yeast cells (Köhler et al. 2017). Humans with compromised immune systems, particularly those suffering from AIDS as a result of having been infected by HIV, are more likely to be infected by this fungus (Azie et al. 2012; Brown et al. 2012a, b). Some regions of the world saw a decrease in the prevalence of cryptococcosis with the advent of antiretroviral medications for HIV patients. In limited supplies areas, where HIV prevalence is high and access to medical services is restricted, there has not been much of an impact. Roughly 220,000 occurrences and over 180,000 fatalities worldwide occur yearly as a result of Cryptococcus infections that move to the brain and cause cryptococcal meningitis (Brown et al. 2012a, b).

Youth meningitis is most commonly caused by Cryptococcus in sub-Saharan Africa. Contrary to *C. neoformans*, *C. gattii*, cryptococcosis is less common but nevertheless occurs in generally healthy individuals; however, those with compromised immune systems or lung disorders are more susceptible. In tropical and subtropical regions, *C. gattii* lives in soil and in collaboration with certain trees. Human and animal infections in British Columbia and the U.S. Pacific Northwest have been linked to *C. gattii* since the late 1990s (Datta and Lal 2012).

## Histoplasma capsulatum

The systemic fungal infection known as histoplasmosis is brought on by *H. capsuleatuim* var. *capslulatlun*. It is the most prevalent respiratory mycotic illness infecting people as well as animals, with a global distribution (Schwarz 1981). Heat-dimorphic fungus *H. capsulatum* thrives in nature as hyphae but in mammals it takes the shape of budding yeast. Histoplasma, the causative agent of *H. capsulatum*, shares diagnostic and pathogenic characteristics with tuberculosis (Kim 2016; Woods 2016).There are many different symptoms associated with histoplasmosis, and the underlying mechanisms and strain specificity of the illness are not well understood. The environmental and genetic factors that contribute to a "successful" infection are not fully understood, but the lack of host defence mechanisms responsible for the intracellular fatal injuries of fungi has made gradually circulated histoplasmosis an increasingly significant opportunistic infection (Wheat et al. 1985).

Initial investigation on *H. capsuilatuii* has drawn more attention recently, partly because of the rising incidence of these infections in people with compromised immune systems or other disabilities, especially since the development of acquired immune deficiency syndrome. Patients with acquired immune deficiency syndrome frequently catch histoplasmosis in locations where the disease is endemic (Johnson et al. 1986), but because of population movement, this condition is becoming more of a problem in non-endemic areas (Huang et al. 1987).

*H. capsiilatiim* is the dimorphic fungal species that has been investigated the most from a biochemical and molecular perspective. It has two phases: a saprobic mycelial phase and a parasitic phase that consists of yeast cells. Yeast cells have an oval shape, measure 1–3 pm in diameter, are found inside macrophages and reticuloendothelial cells, and are cultured at 37 °C. Yeast cells divide via budding while they are in culture, and un-budded cells have a single nucleus (Edwards et al. 1959).

Under unsuitable circumstances, the hyphae form two distinct kinds of conidia: microconidia (Edwards et al. 1960; Howard 1962) and macroconidia (Conant et al. 1971; Goos 1964). Microconidia are the preferred infectious form of *H. capsiilatum*, most likely because of their microscopic dimension. At 37 °C, polar or nonpolar budding may stimulate microconidia, which contain several exterior tubercules, to emerge and generate yeast cells (Garrison and Boyd 1977, 1978; Pine and Webster 1962). The one and only instance of the organism found in the infested tissue is yeast cells.

## Pneumocystis jirovecii

*Pneumocystis jirovecii* is an opportunistic fungal pathogen that causes severe lung disease to patients at risk. *Pneumocystis pneumonia* (PCP), one of the most common and serious opportunistic infections in immunocompromised people, is caused by *Pneumocystis jirovecii* (previously known as *Pneumocystis carinii* f. sp. *hominis*) (Stringer et al. 2002). Pneumocystis organisms are an extensive group of atypical fungi that are widely distributed and exhibit lung tropism. Notably, each species of Pneumocystis has a significant affinity for a particular species of mammalian host (Dei-Cas 2000; Gigliotti et al. 1993).

According to serologic investigations, the majority of children exhibit early detection of specific antibodies to the pathogen (Medrano et al. 2003; Meuwissen et al. 1977; Pifer et al. 1978), indicating regular contact to this organism. This discovery has led to the long-held belief that the disease in immunocompromised individuals arises from the revival of a repressed infection that initially developed during a young age. Pneumocystis has also been discovered in immunocompromised individuals not using PCP (Nevez et al. 1999). The development of increasingly sensitive techniques has made it possible to identify Pneumocystis in healthy, normal individuals. Immunofluorescence labelling and polymerase chain reaction (PCR) have made it possible to identify pneumocystis in respiratory samples collected using non-invasive techniques (Eisen et al. 1994; Helweg-Larsen et al. 1998).

## Diagnostic Test for Fungal Infection

Fungal infections are usually diagnosed by an approach that consists of laboratory testing, diagnostic imaging, and clinical examination. Here is an overview of typical diagnostic techniques:

## *Non-molecular Methods*

Non-molecular diagnostic methods are conventional test methods, which include whole blood, serum, plasma, and urine test. For the diagnosis of invasive fungal infections, non-molecular diagnostic methods continue to be preferred (Arvanitis et al. 2014).

### Clinical Assessment

To estimate the probability of a fungal infection, specialists evaluate signs and symptoms, health records, and associated risks (such as immunosuppression or recent antibiotic usage).

### Microscopic Examination

Fungal elements can be identified through direct microscopy of clinical specimens (such as skin scrapings, sputum, blood, or urine) using stains like potassium hydroxide (KOH) or calcofluor white.

### Culture Identification

Fungi from clinical samples are isolated in specialized media. This aids in identifying the precise species and figuring out their sensitivity to antifungals.

### Antigen and Antibody Assays

Antigen and Antibody assays, are widely used in fungal infection diagnosis. Aspergillosis and cryptococcosis are two major examples of fungal infections that can be identified with the use of particular antigen or antibody testing like β-D-glucan assay, Galactomannan (GM) assay. Here both antigen-antibiotic assay are outlined briefly.

#### β-D-Glucan Assay

The β-D-glucan assay is often useful in combination with culture. Overall, the sensitivities of β-D-glucan testing in individual studies have ranged from 55% to 95%, and specificities from 77% to 96%, for patients with hematologic malignancies who are suffering from invasive aspergillosis (Kawazu et al. 2004; Koo et al. 2009; Obayashi et al. 2008; Odabasi et al. 2004; Ostrosky-Zeichner et al. 2005; Pickering et al. 2005).

Galactomannan (GM) Assay

The galactomannan (GM) assay is a fairly specific and sensitive test for the diagnosis of invasive aspergillosis, infection caused by *Histoplasma capsulatum* and *Fusarium* spp. can be performed in serum, BAL fluid, CSF, or pleural fluid (Arvanitis et al. 2014).

#### Diagnostic Imaging Techniques

To find fungal infections in organs or tissues, imaging diagnostic tools, such as CT scans, MRIs, or chest X-rays might be utilized. When invasive fungal infections damage deep tissues or organs, this is especially helpful.

#### Biopsy

A specimen biopsy could be required in certain circumstances in order to collect tissue samples for a histopathological analysis. This can offer unmistakable proof of a fungal infection and support treatment decisions. The suspected location and kind of fungal infection influence the diagnostic test selection. To get an appropriate diagnosis, these techniques are frequently combined.

### *Molecular Methods*

Molecular techniques, prominent among them the use of PCR, have supplanted conventional diagnostic procedures for a wide range of human illnesses and are employed on a daily basis in ordinary clinical practice (Strick and Wald 2006). Among their primary advantages over conventional approaches are their simplicity, convenience of usage, and quick turnaround time. Consequently, it is not unexpected that these techniques have been highlighted for years as a possible remedy for the IFI diagnosis issue (Rüchel 1993).

#### PCR

One of the most widely established and popular molecular approaches for fungal diagnoses is PCR. Molecular methods such as polymerase chain reaction (PCR) can be used to acquire fungal DNA from clinical specimens with high specificity and sensitivity. Over the course of time, numerous PCR tests for invasive *Aspergillus* sp. infections have been created. Their unique characteristics and concerns, however, vary widely in the clinical data (Buchheidt et al. 2001; Guinea et al. 2013; Kawazu et al. 2004; Musher et al. 2004; Reinwald et al. 2013; Springer et al. 2011), ranging from 43% to 100% and 64% to 100%, respectively.

### Multiplex PCR

A potential remedy for this issue would be to employ the SeptiFast assay, a commercially manufactured broad-range multiplex PCR that can identify a wide variety of bacteria and fungi and is now being used to identify the pathogen in sepsis cases (Lamoth et al. 2010; Lucignano et al. 2011). Similarly, the Film-Array system is a series of assays that has just been developed and commercialized for use in the microbiology laboratory. It can quickly identify >25 pathogens, especially many *Candida* spp., and antibiotic resistance genes in positive samples of blood within 1 h (Blaschke et al. 2012). Lastly, broad-range PCR techniques may be employed to quickly identify the pathogen during outbreak situations.

### Novel Molecular Method

Over the years, a wide range of additional molecular techniques (Table 6) have been developed and evaluated, either as stand-alone tests or as a supplement to enhance the efficiency of PCR. For instance, fluorescence in situ hybridization (FISH) is a method that finds specific regions on the genomes of microbial pathogens in human samples using fluorescent probes. These regions may then be identified using fluorescence microscopy.

This technique has been shown to have excellent accuracy for the detection of Candida sp. infestations from blood sample bottles and has been used as a supplement to culture (Wilson et al. 2005) or PCR (Rickerts et al. 2011). Moreover, results from two investigations on invasive fungal rhinosinusitis (Montone et al. 2010) and coccidioidomycosis (Montone et al. 2011) demonstrate the method's excellent efficacy on freeze-dried tissue sections, even in situations when cultures are unavailable or have not been carried out. While RNA polymerase is used to amplify RNA rather than DNA, nucleic acid sequence-based amplification (NASBA) is an isothermal approach that is remarkably similar to PCR (Compton 1991). Using mass spectrometry as its foundation, MALDI-TOF MS is an alternative method for identifying the distinct protein fingerprints of various bacteria. It is feasible to identify the observed pathogen at the genus, species, and even strain levels by directly comparing the organism's spectral pattern with databases of known patterns from various microbes (Hettick et al. 2008).

## Treatment and Management of Fungal Infection

The type of fungus and the affected body part determine the course of treatment for fungal diseases. While more serious fungal infections might be treated for up to a year, less prevalent diseases can be cleared up with short-term measures. Depending on the nature and extent of the disease, antifungal medications either external or oral may be used to treat fungal infections. Fungal infections can be prevented by

**Table 6** Technology behind novel diagnostic methods for fungal infections

| Novel molecular Methods | Methodology | References |
|---|---|---|
| FISH (fluorescent in situ hybridization) | Application of fluorescent probes combined with a specific target sequence are assorted with the tested sample | Rickerts et al. (2011), Wilson et al. (2005) |
| Nucleic acid sequence-based amplification (NASBA) | Analogous to PCR but varies in the sense that it enlarges RNA by using an RNA polymerase, additionally it is isothermal. | Compton (1991), Loeffler et al. (2001) |
| MALDI-TOF (matrix-assisted laser description ionization time–of –flight mass spectrometry) | A test sample is poured into a matrix material-filled hole that has the capacity to absorb UV radiation and turn it to heat. The mixture is the target of a laser beam. A cloud of ionized proteins and matrix is produced when the laser beam is absorbed by the matrix and some of the analyte-matrix mixture vaporizes and becomes ionized. Particles located in this cloud eventually move in the direction of a sensor when they are exposed to an electric field. Each particle's mass and charge dictate how long it takes for it to reach the detector. This enables the mass spectrometer to ascertain the properties of the particles contained in the sample under test. The microorganism in the sample can be identified by comparing the generated spectral pattern to a standard database. | De Carolis et al. (2012), Hettick et al. (2008) |
| SERRS (surface enhanced resonance Raman spectroscopy) | The specimen sample is positioned on an uneven surface, contributing to the dispersed lighting effect. The material is supplemented with light-emitting dyes and DNA probes. Following the probes' binding to the target DNA in the sample, a double-stranded DNA exonuclease is introduced to the mixture and begins to digest all of the bound probes, leaving the open, single-stranded probes undigested. Ultimately, a sensor picks up and analyses the dispersed light from the undigested probe, revealing the digested sensors' genetic sequence. | Faulds et al. (2005) |
| T2 nuclear magnetic resonance (T2NMR) | Nuclear magnetic resonance (NMR) T2. A microorganism identified in the sample has its desired sequence expanded first. The expanded sequence is then allowed to hybridize with DNA probes that are attached to paramagnetic nanoparticles and added to the amplicon. As a result, the target is identified by magnetic resonance imaging, which detects the shift in the nanoparticles' T2 relaxation time. | Neely et al. (2013) |

practicing proper hygiene, not sharing personal objects, and maintaining dry and clean skin. Research on human fungal pathogens will continue to expand, and important findings in the near future will aid in the creation of fresh tactics to fight deadly fungal infections (Kim 2016).

The challenge of discovering effective ways to kill the fungi specifically is a limiting issue in the discovery of novel antifungal medicines. The basics of the molecular as well as cell biology of fungi is incredibly analogous to the biology of animals, in contrast to bacterial pathogens. Fungal infestation of the skin and keratinized tissues are curable and easily identified, by the help of physical examination, patient history, and microscopic and culture analysis of specimens (skin or nail). The majority of patients have moderate to non-life-threatening symptoms, although the effects can still be quite detrimental to the individual's level of life (Drake et al. 1998; Drake and Scher 1995; Finlay 1997; Morgan et al. 1997). A condition like toenail dystrophy in onychomycosis can make it difficult to walk, stand, exercise, or wear shoes correctly, and nail infections can make it difficult to dress or type. People suffering with unattractive nail infections may also feel ashamed of the disease, which can negatively impact their social and personal interactions by lowering their self-confidence and sense of self-worth (Drake et al. 1998; Lubeck et al. 1993; Scher 1994).

Infestations caused by superficial fungi, especially those affecting the toenails and feet, can also serve as a source of microorganisms that are transmissible to other parts of one's body or to other people. Patients with impaired immune systems in particular need to be monitored closely for an early indication of a higher chance of these infections progressing to an invasive form. Treatment for patients with skin disorders that are not life-threatening, but cause disfigurement is therefore justified.

For instance, ergosterol, a crucial component of the fungal cell membrane that resembles cholesterol in mammals, is targeted by two of the present medication classes. Finding medications that destroy the fungus without having major adverse effects on individuals is difficult because of the fundamental similarities between the fungal and mammalian kingdoms. It's a battle against time as researchers explore for novel and creative ways to counteract fungal infections, since fungus can quickly develop resistance to the medications we use to eradicate them. Indeed, multidrug-resistant bacteria are currently proliferating globally, and clinical resistance to every class of antifungal medication has surfaced (Gigliotti et al. 1993).

The clinical appearance may resemble illnesses brought on by mycobacteria and other infectious agents, delaying diagnosis (Garber 2001). Whenever sclerotic bodies are observed, their existence is confirmed by the histological study of scrapings or biopsy material. This condition has no known cure; however many therapy modalities have shown promise (Bonifaz et al. 1997). In persistent mucocutaneous candidiasis, a rare disorder, salivary glands are persistently infected (typically by *Candida albicans*), and the microbe's infection can spread to the skin and nails. Though the underlying deficiency is yet unclear, the disorder is linked to compromised cell-mediated responses to Candida (Garber 2001; Lilic et al. 1996).

Adaptive cell-mediated immunity (CMI) is essential for eliminating the pathogen *C. neoformans*, mainly those with infection suffer from immune suppression as a result of organ donation or other immunological disorders, which prevents adaptive CMI from working effectively (Hole and Wormley 2016). Therefore, it is necessary to comprehend the manner in which the body's natural immunity fights against *C. neoformans* in order to create effective medicines to assist these individuals recover from the cryptococcosis. Functions of phagocytic cells, including natural killer cells, dendritic cells, and macrophages, together with the molecular mechanisms underlying innate immunity play important role in phagocytosing and eliminating fungal infections like *C. neoformans* (Hole and Wormley 2016; Kim 2016). The following are some broad fungal disease management principles:

### *Antifungal Medications*

Antifungal drugs are the mainstay of treatment for the majority of fungal infections. Antifungal medications fall into various classes, such as azoles, polyenes, echinocandins, and allylamines. The type of fungus and the infection site determine which medicine is best.

### *Topical Antifungals*

Topical antifungal creams or ointments are frequently useful for treating superficial fungal infections, such as athlete's foot or cutaneous yeast infections.

### *Systemic Antifungals*

Systemic antifungal medication, either intravenously or orally, is typically necessary for invasive fungal infections or those affecting internal organs.

### *Surgical Intervention*

Surgery can be required in certain situations to empty fungal-induced abscesses or remove affected tissue.

### *Supportive Care*

Patients who have serious fungal infections could need supportive treatment, which includes nutritional assistance, intravenous fluids, and problem management.

### *Antifungal Prophylaxis*

Antifungal agents may be administered in advance to avoid fungal infections in some high-risk individuals, such as those receiving chemotherapy or undergoing bone marrow transplantation.

### *Management of Underlying Conditions*

Treating underlying conditions that predispose individuals to fungal infections, such as diabetes or immunosuppression, is essential to prevent recurrent infections.

### *Monitoring*

Patients on antifungal therapy should be routinely monitored in order to evaluate response to treatment, keep an eye out for side effects, and modify the medication as necessary. It is imperative that medical professionals customise the management strategy to the particular fungal infection and the unique conditions of each patient. Improving outcomes for individuals with fungal infections requires prompt and adequate treatment.

## Conclusion

Human fungal infections, both systemic and local, are becoming more common at a startling rate. This is mostly because of improvements in medical practice, which have led to an increase in the number of hospitalized patients who are critically ill and immunocompromised. This rising pool of people at risk has been made more so by the HIV epidemic and other immune system disorders. Emerging pathogens, such as *C. auris*, which is frequently multi-resistant, challenging to identify, and has recently caused outbreaks in hospital settings, are a developing concern.

To reduce the chance of develop, it is generally preferable to treat localized fungal infections. A substantial number of deaths are linked to systemic fungal infections in immunocompromised patients, including recipients of solid organ transplants and bone marrow. The frequency of fungal infections can be decreased within medical centers by reducing risk factors, such as inadequate cleanliness practices, but the necessity for appropriate therapies is underscored by individual's incapacity to consistently prevent such infections.

There is a serious risk to society from newly emerging fungal infections, especially those for which there are now no viable treatments. Conventional methods have comparatively limited sensitivity and often cause significant delays in diagnosis and initiation of appropriate treatment; therefore, advanced diagnostic approaches, such as molecular- or bioassaybased methods, should be adopted in routine practice. Research and infestation awareness programme on human–fungal interaction or human host–pathogen interaction should keep growing. To create cutting-edge therapies for deadly fungal diseases, new information and discoveries should be applied.

## References

Alcamo IE (2003) Microbes and society: an introduction to microbiology. Jones & Bartlett Learning, Burlington

Alexopoulos CJ, Ahmadjian V, Moore D (2024) Fungus. Encycl Br. https://www.britannica.com/science/fungus. Accessed 26 Apr 2024

Almeida F, Wolf JM, Casadevall A (2015) Virulence-associated enzymes of *Cryptococcus neoformans*. Eukaryot Cell 14:1173–1185

Anonymous (2017) Stop neglecting fungi. Nat Microbiol 2:17120. https://doi.org/10.1038/nmicrobiol.2017.120

Anonymous (2024) The fungi. ATW public domain. https://web.archive.org/web/20120620205340/http:/palaeos.com/fungi/fungi.html. Accessed 1 May 2024

Arvanitis M, Anagnostou T, Fuchs BB, Caliendo AM, Mylonakis E (2014) Molecular and non-molecular diagnostic methods for invasive fungal infections. Clin Microbiol Rev 27:490–526. https://doi.org/10.1128/CMR.00091-13

Azie N, Neofytos D, Pfaller M, Meier-Kriesche H-U, Quan S-P et al (2012) The PATH (Prospective Antifungal Therapy) Alliance® registry and invasive fungal infections: update 2012. Diagn Microbiol Infect Dis 73:293–300. https://doi.org/10.1016/j.diagmicrobio.2012.06.012

Bahn YS, Jung KW (2013) Stress signaling pathways for the pathogenicity of *Cryptococcus*. Eukaryot Cell 12:1564–1577

Berkeley (2024) Introduction to the fungi. Virtual Libr. https://ucmp.berkeley.edu/fungi/fungi.html. Accessed 29 Apr 2024

Blaschke AJ, Heyrend C, Byington CL, Fisher MA, Barker E et al (2012) Rapid identification of pathogens from positive blood cultures by multiplex polymerase chain reaction using the FilmArray system. Diagn Microbiol Infect Dis 74:349–355. https://doi.org/10.1016/j.diagmicrobio.2012.08.013

Bonifaz A, Martinez-Soto E, Carrasco-Gerard P, J. (1997) Treatment of chromoblastomycosis with itraconazole, cryosurgery, and a combination of both. Int J Dermatol 36:542–547

Bozkurt TO, Kamoun S (2020) The plant–pathogen haustorial interface at a glance. J Cell Sci 133. https://doi.org/10.1242/jcs.237958

Britannica (2024) Parasitism in humans. Encycl Br. https://www.britannica.com/science/fungus/Parasitism-in-humans. Accessed 28 Apr 2024

Bronson DM, Desai DR, Barsky S, Foley SM (1983) An epidemic of infectionwith *Trichophyton tonsurans* revealed in a 20-year survey of fungal infections in Chicago. J Am Acad Dermatol 8:322–330

Brown GD, Denning DW, Gow NAR, Levitz SM et al (2012a) Hidden killers: human fungal infections. Sci Transl Med 4. https://doi.org/10.1126/scitranslmed.3004404

Brown GD, Denning DW, Levitz SM (2012b) Tackling human fungal infections. Science 336:647–647. https://doi.org/10.1126/science.1222236

Buchheidt D, Baust C, Skladny H, Ritter J, Suedhoff T et al (2001) Detection of aspergillus species in blood and Bronchoalveolar lavage samples from immunocompromised patients by means

of 2-step polymerase chain reaction: clinical results. Clin Infect Dis 33:428–435. https://doi.org/10.1086/321887
Casadevall A, Cordero RJB, Bryan R, Nosanchuk J, Dadachova E (2017) Melanin, radiation, and energy transduction in fungi. Microbiol Spectr 5:0037
Chang ST, Miles PG (2004) Mushrooms: cultivation, nutritional value, medicinal effect and environmental impact. CRC Press, Boca Raton
Chapman SW, Daniel CR 3rd. (1994) Cutaneous manifestations of fungal infection. Infect Dis Clin N Am 8:879–910
Cheek M, Nic Lughadha E, Kirk P, Lindon H, Carretero J et al (2020) New scientific discoveries: plants and fungi. Plants People Planet 2:371–388. https://doi.org/10.1002/ppp3.10148
Chiritescu MM, Chiritescu ME, Scher RK (1996) Newer systemic antifungal drugs for the treatment of onychomycosis. Clin Podiatr Med Surg 13:741–758
Choi J, Jung WH, Kronstad JW (2015) The cAMP/protein kinase A signaling pathway in pathogenic basidiomycete fungi: connections with iron homeostasis. J Microbiol 53:579–587
Compton J (1991) Nucleic acid sequence-based amplification. Nature 350:91–92. https://doi.org/10.1038/350091a0
Conant NF, Smith DT, Baker RD, Callaway JL (1971) Manual of clinical mycology. The W.B. Saunders Co., Philadelphia
Dagenais TRT, Keller NP (2009) Pathogenesis of *Aspergillus fumigatus* in invasive Aspergillosis. Clin Microbiol Rev 22:447–465. https://doi.org/10.1128/CMR.00055-08
Datta J, Lal N (2012) Application of molecular markers for genetic discrimination of fusarium wilt pathogen races affecting chickpea and pigeonpea in major regions of India. Cell Mol Biol 58:55–65. https://doi.org/10.1170/T921
De Carolis E, Vella A, Florio AR, Posteraro P, Perlin DS et al (2012) Use of matrix-assisted laser desorption ionization–time of flight mass spectrometry for caspofungin susceptibility testing of *Candida* and *Aspergillus* species. J Clin Microbiol 50:2479–2483. https://doi.org/10.1128/JCM.00224-12
Deacon J (2005) Fungal biology. Blackwell Publishers, Cambridge
Dei-Cas E (2000) Pneumocystis infections: the iceberg? Med Mycol 38:23–32
Denning DW, Evans EGV, Kibbler CC, Richardson MD, Roberts MM et al (1997) Guidelines for the investigation of invasive fungal infections in haematological malignancy and solid organ transplantation. Eur J Clin Microbiol Infect Dis 16:424–436
Deorukhkar SC, Saini S (2015) Candidiasis: past, present and future. Int J Infect Trop Dis 2:12–24
Detandt M, Nolard N (1995) Fungal contamination of the floors of swimming pools, particularly subtropical swimming paradises. Mycoses 38:509
Diz-Dios P, Ocampo A, Miralles C, Otero I, Iglesias I et al (1999) Frequency of oropharyngeal candidiasis in HIV-infected patients on protease inhibitor therapy. Oral Surg Oral Med Oral Pathol Oral Radiol Endod 87:437–441
Drake LA, Scher RK (1995) Onychomycosis: a significant and important disease. In: Second international symposium on onychomycosis, Florence, Italy
Drake LA, Scher RK, Smith EB, Faich GA, Smith SL et al (1998) Effect of onychomycosis on quality of life. J Am Acad Dermatol 38:702–704
Edwards MR, Hazen EL, Edwards GA (1959) The fine structures of the yeast-like cells of Histoplasma capsula-tilminculture. J Gen Microbiol 20:496–503
Edwards MR, Hazen EL, Edwards GA (1960) The micromorphology of the tuberculate spores of Histoplasma 1b. Can J Microbiol 6:65–70
Eisen D, Ross BC, Fairbairn J, Warren RJ, Baird RW et al (1994) Comparison of *Pneumocystis carinii* detection by toluidine blue O staining, direct immunofluorescence and DNA amplification in sputum specimens from HIV patients. Pathology 26:198–200
Elewski BE (2000) Tinea capitis: a current perspective. J Am Acad Dermatol 42:1–20
Ellis DH, Watson AB, Marley JE, Williams TG (1997) Non-dermatophytes in onychomycosis of the toenails. Br J Dermatol 136:490–493
Evans EG (1997) Tinea pedis: clinical experience and efficacy of short treatment. Dermatology 194:3–6

Faulds K, Smith WE, Graham D (2005) DNA detection by surface enhanced resonance Raman scattering (SERRS). Analyst 130:1125. https://doi.org/10.1039/b500248f

Finlay AY (1997) Quality of life measurement in dermatology: a practical guide. Br J Dermatol 136:305–314

Fisher MC, Henk DA, Briggs CJ, Brownstein JS, Madoff LC et al (2012) Emerging fungal threats to animal, plant and ecosystem health. Nature 484:186–194. https://doi.org/10.1038/nature10947

Fisher MC, Gow NAR, Gurr SJ (2016) Tackling emerging fungal threats to animal health, food security and ecosystem resilience. Philos Trans R Soc B Biol Sci 371:20160332. https://doi.org/10.1098/rstb.2016.0332

Fisher MC, Hawkins NJ, Sanglard D, Gurr SJ (2018) Worldwide emergence of resistance to antifungal drugs challenges human health and food security. Science 360:739–742. https://doi.org/10.1126/science.aap7999

Garber G (2001) An overview of fungal infections. Drugs 61:1–12

Garrison RG, Boyd KS (1977) The fine structure of microconidi'al germination and vegetative cells of Histoplasnacapsliatumi. Ann Microbiol 128:137–149

Garrison R, Boyd KS (1978) Electron microscopy of yeast-like cells development from the microconidilim of Histoplasma capsulatum. J Bacteriol 133:345–353

Gigliotti F, Harmsen AG, Haidaris CG, Haidaris PJ (1993) *Pneumocystis carinii* is not universally transmissible between mammalian species. Infect Immun 61:2886–2890

Goos RD (1964) Germination of the macroconidium of Histoplasma capsulatum. Mycologia 5:662–671

Greer DL (1995) Evolving role of nondermatophytes in onychomycosis [review]. Int J Dermatol 34:521–524

Groll AH, Shah PM, Mentzel C, Schneider M, Just-Nuebling G (1996) Trends in the post-mortem epidemiology of invasive fungal infections at a university hospital. J Infect 33:23–32

Guarro J, Gené J, Stchigel AM (1999) Developments in fungal taxonomy. Clin Microbiol Rev 12:454–500. https://doi.org/10.1128/CMR.12.3.454

Guinea J, Padilla C, Escribano P, Muñoz P, Padilla B et al (2013) Evaluation of MycAssay™ *Aspergillus* for diagnosis of invasive pulmonary Aspergillosis in patients without hematological cancer. PLoS One 8:e61545. https://doi.org/10.1371/journal.pone.0061545

Gupta AK, Konnikov N, MacDonald P, Rodger NW, Edmonds MW et al (1998) Prevalence andepidemiology of toenail onychomycosis in diabetic subjects: a multicentre survey. Br J Dermatol 139:665–671

Hanson JR (2008) The chemistry of fungi. Royal Society of Chemistry, London

Harris SD (2008) Branching of fungal hyphae: regulation, mechanisms and comparison with other branching systems. Mycologia 100:823–832. https://doi.org/10.3852/08-177

Hawksworth DL, Lücking R (2017) Fungal diversity revisited: 2.2 to 3.8 million species. Microbiol Spectr 5:79–95. https://doi.org/10.1128/microbiolspec.FUNK-0052-2016

Hay RJ (2006) Fungal infections. Clin Dermatol 24:201–212. https://doi.org/10.1016/j.clindermatol.2005.11.011

Helweg-Larsen J, Jensen JS, Benfield T, Svendsen UG, Lundgren JD et al (1998) Diagnostic use of PCR for detection of *Pneumocystis carinii* in oral wash samples. J Clin Microbiol 36:2068–2072

Hettick JM, Green BJ, Buskirk AD, Kashon ML, Slaven JE et al (2008) Discrimination of *aspergillus* isolates at the species and strain level by matrix-assisted laser desorption/ionization time-of-flight mass spectrometry fingerprinting. Anal Biochem 380:276–281. https://doi.org/10.1016/j.ab.2008.05.051

Hibbett DS, Binder M, Bischoff JF, Blackwell M, Cannon PF et al (2007) A higher-level phylogenetic classification of the fungi. Mycol Res 111:509–547. https://doi.org/10.1016/j.mycres.2007.03.004

Hoepelman IM, Dupont B (1996) Oral candidiasis: the clinical challenge of resistance and management. Int J Antimicrob Agents 6:155–159

Hole C, Wormley FL (2016) Innate host defense against *Crytococcus neoformans*. J Microbiol 54:202–211
Howard DH (1962) The morphogenesis of the parasitic formofdimorphic fungi. Mycopathologia 18:127–139
Huang CT, McGarry T, Cooper S, Saunders R, Andavolu R (1987) Disseminated histoplasmosis in the acquliled immuno deficiency syndrome. Arch Intern Med 47:1181–1184
Jeffery-Smith A, Taori SK, Schelenz S, Jeffery K, Johnson EM et al (2018) *Candida auris*: a review of the literature. Clin Microbiol Rev 31. https://doi.org/10.1128/CMR.00029-17
Johnson PC, Sarosi GA, Septimus EJ, Satterwhite IK (1986) Progressive disseminated histoplasmosis in patients with the acquired immune deficiency syndrome: a report of 12 cases and a literature review. Semin Respir Infect 1:1–8
Kawazu M, Kanda Y, Nannya Y, Aoki K, Kurokawa M et al (2004) Prospective comparison of the diagnostic potential of real-time PCR, double-sandwich enzyme-linked immunosorbent assay for galactomannan, and a (1→3)-β-d-glucan test in weekly screening for invasive Aspergillosis in patients with hematological disorders. J Clin Microbiol 42:2733–2741. https://doi.org/10.1128/JCM.42.6.2733-2741.2004
Kim JY (2016) Human fungal pathogens: why should we learn? J Microbiol 54:145–148. https://doi.org/10.1007/s12275-016-0647-8
Köhler JR, Hube B, Puccia R, Casadevall A, Perfect JR (2017) Fungi that infect humans. Microbiol Spectr 5. https://doi.org/10.1128/microbiolspec.FUNK-0014-2016
Konopka JB, Casadevall A, Taylor JW, Heitman J, Cowen L (2019) One health: fungal pathogens of humans, animals, and plants. Am Acad Microbiol. https://doi.org/10.1128/AAMCol.18Oct.2017
Koo S, Bryar JM, Page JH, Baden LR, Marty FM (2009) Diagnostic performance of the (1→3)-beta-D-glucan assay for invasive fungal disease. Clin Infect Dis 49:1650–1659. https://doi.org/10.1086/647942
Kullberg BJ, Arendrup MC (2015) Invasive candidiasis. N Engl J Med 373:1445–1456. https://doi.org/10.1056/NEJMra1315399
Lamoth F, Jaton K, Prod'hom G, Senn L, Bille J et al (2010) Multiplex blood PCR in combination with blood cultures for improvement of microbiological documentation of infection in febrile neutropenia. J Clin Microbiol 48:3510–3516. https://doi.org/10.1128/JCM.00147-10
Levy LA (1997) Epidemiology of onychomycosis in special-risk populations. J Am Podiatr Med Assoc 87:546–550
Li Y, Sun R, Yu J, Saravanakumar K, Chen J (2016) Antagonistic and biocontrol potential of *Trichoderma asperellum* ZJSX5003 against the maize stalk rot pathogen *Fusarium graminearum*. Indian J Microbiol 56:318–327. https://doi.org/10.1007/s12088-016-0581-9
Lilic D, Cant AJ, Abinun M, Calvert JE, Spickett GP (1996) Chronic mucocutaneous candidiasis. I. Altered antigen-stimulated IL-2, IL-4, IL-6 and interferon-gamma (IFN-gamma) production. Clin Exp Immunol 105:205–212
Loeffler J, Hebart H, Cox P, Flues N, Schumacher U et al (2001) Nucleic acid sequence-based amplification of aspergillus RNA in blood samples. J Clin Microbiol 39:1626–1629. https://doi.org/10.1128/JCM.39.4.1626-1629.2001
Lubeck DP, Patrick DL, McNulty P, Fifer SK, Birnbaum J (1993) Quality of life of persons with onychomycosis. Qual Life Res 2:341–348
Lucignano B, Ranno S, Liesenfeld O, Pizzorno B, Putignani L et al (2011) Multiplex PCR allows rapid and accurate diagnosis of bloodstream infections in newborns and children with suspected sepsis. J Clin Microbiol 49:2252–2258. https://doi.org/10.1128/JCM.02460-10
Medrano FJ, Respaldiza N, Medrano A, Varela JM, Montes-Cano M et al (2003) Seroprevalence of pneumocystis human infection in southern Spain. J Eukaryot Microbiol 50:649–650
Mendling W, Gutschmidt J, Gantenberg R, Andrade MP, Schoenian G (1998) Comparison of strain specificity of yeasts from various organs of women with vaginal candidiasis [in German]. Mycoses 41:23–25

Meuwissen JT, Tauber I, Leeuwenberg A, Beckers P, Sieben M (1977) Parasitologic and serologic observations of infection with pneumocystis in humans. J Infect Dis 136:43–49
Money NP (2016) Fungal diversity. In: The fungi, 3rd edn. Elsevier Ltd, Amsterdam. https://doi.org/10.1016/B978-0-12-382034-1.00001-3
Montone KT, Litzky LA, Feldman MD, Peterman H, Mathis B et al (2010) In situ hybridization for Coccidioides immitis 5.8S ribosomal RNA sequences in formalin-fixed, paraffin-embedded pulmonary specimens using a locked nucleic acid probe. Diagn Mol Pathol 19:99–104. https://doi.org/10.1097/PDM.0b013e3181b3aa55
Montone KT, LiVolsi VA, Lanza DC, Kennedy DW, Palmer J et al (2011) In situ hybridization for specific fungal organisms in acute invasive fungal Rhinosinusitis. Am J Clin Pathol 135:190–199. https://doi.org/10.1309/AJCPQLYZBDF30HTM
Morgan M, McCreedy R, Simpson J, Hay RJ (1997) Dermatology quality of life scales—a measure of the impact of skin diseases. Br J Dermatol 136:202–206
Musher B, Fredricks D, Leisenring W, Balajee SA, Smith C et al (2004) Aspergillus galactomannan enzyme immunoassay and quantitative PCR for diagnosis of invasive Aspergillosis with Bronchoalveolar lavage fluid. J Clin Microbiol 42:5517–5522. https://doi.org/10.1128/JCM.42.12.5517-5522.2004
Neely LA, Audeh M, Phung NA, Min M, Suchocki A et al (2013) T2 magnetic resonance enables nanoparticle-mediated rapid detection of Candidemia in whole blood. Sci Transl Med 5:182ra54. https://doi.org/10.1126/scitranslmed.3005377
Nevez G, Raccurt C, Jounieaux V, Dei-Cas E, Mazars E (1999) Pneumocystosis versus pulmonary pneumocystis carinii colonization in HIV-negative and HIV-positive patients. AIDS 13:535–556
O'Meara TR, Alspaugh JA (2012) The Cryptococcus neoformans capsule: a sword and a shield. Clin Microbiol Rev 25:387–408
Obayashi T, Negishi K, Suzuki T, Funata N (2008) Reappraisal of the serum (1→3)-β-D-Glucan assay for the diagnosis of invasive fungal infections—a study based on autopsy cases from 6 years. Clin Infect Dis 46:1864–1870. https://doi.org/10.1086/588295
Odabasi Z, Mattiuzzi G, Estey E, Kantarjian H, Saeki F et al (2004) Beta-D-Glucan as a diagnostic adjunct for invasive fungal infections: validation, cutoff development, and performance in patients with acute Myelogenous leukemia and myelodysplastic syndrome. Clin Infect Dis 39:199–205. https://doi.org/10.1086/421944
Ostrosky-Zeichner L, Alexander BD, Kett DH, Vazquez J, Pappas PG et al (2005) Multicenter clinical evaluation of the (1→3)β-D-Glucan assay as an aid to diagnosis of fungal infections in humans. Clin Infect Dis 41:654–659. https://doi.org/10.1086/432470
Panackal AA, Wuest SC, Lin YC, Wu T, Zhang N et al (2015) Paradoxical immune responses in non-HIV Cryptococcal meningitis. PLoS Pathog 11:E1004884
Parniske M (2008) Arbuscular mycorrhiza: the mother of plant root endosymbioses. Nat Rev Microbiol 6:763–775. https://doi.org/10.1038/nrmicro1987
Pfaller MA, Diekema DJ (2010) Epidemiology of invasive mycoses in North America. Crit Rev Microbiol 36:1–53. https://doi.org/10.3109/10408410903241444
Pickering JW, Sant HW, Bowles CAP, Roberts WL, Woods GL (2005) Evaluation of a (1→3)-β-d-Glucan assay for diagnosis of invasive fungal infections. J Clin Microbiol 43:5957–5962. https://doi.org/10.1128/JCM.43.12.5957-5962.2005
Piérard GE, Arrese JE, Piérard-Franchimont C (1996) Treatment and prophylaxis of tinea infections. Drugs 52:209–224
Pifer LL, Hughes WT, Stagno S, Woods D (1978) Pneumocystis carinii infection: evidence for high prevalence in normal and immunosuppressed children. Pediatrics 61:35–41
Pine L, Webster RE (1962) Conversion in strains of Histoplasma capsulatum. J Bacteriol 83:149–157
Powderly WG, Mayer KH, Perfect JR (1999) Diagnosis and treatment of oropharyngeal candidiasis in patients infected with HIV: a critical reassessment. AIDS Res Hum Retrovir 15:1405–1412

Rautemaa-Richardson R, Richardson MD (2017) Systemic fungal infections. Medicine (Baltimore) 45:757–762. https://doi.org/10.1016/j.mpmed.2017.09.007
Regulez P, Garcia-Fernandez JF, Moragues MD, Schneider J, Quindos G et al (1994) Detection of anti-Candida albicans IgE antibodies in vaginal washes from patients with acute vulvovaginal candidiasis. Gynecol Obs Invest 37:110–114
Reinwald M, Buchheidt D, Hummel M, Duerken M, Bertz H et al (2013) Diagnostic performance of an aspergillus-specific nested PCR assay in cerebrospinal fluid samples of immunocompromised patients for detection of central nervous system Aspergillosis. PLoS One 8:e56706. https://doi.org/10.1371/journal.pone.0056706
Richardson MD, Warnock DW, Campbell CK (1995) Slide atlas of fungal infection. Blackwell Science Limited, Oxford
Rickerts V, Khot PD, Myerson D, Ko DL, Lambrecht E et al (2011) Comparison of quantitative real time PCR with sequencing and ribosomal RNA-FISH for the identification of fungi in formalin fixed, paraffin-embedded tissue specimens. BMC Infect Dis 11:202. https://doi.org/10.1186/1471-2334-11-202
Rios-Fabra A, Oreno AR, Isturiz RE (1994) Fungal infection in Latin American countries. Infect Dis Clin N Am 8:129–154
Robert VA, Casadevall A (2009) Vertebrate endothermy restricts most fungi as potential pathogens. J Infect Dis 200:1623–1626. https://doi.org/10.1086/644642
Rüchel R (1993) Diagnosis of invasive mycoses in severely immunosuppressed patients. Ann Hematol 67:1–11. https://doi.org/10.1007/BF01709659
Satoh K, Yamazaki T, Nakayama T, Umeda Y, Alshahni et al (2016) Characterization of fungi isolated from the equipment used in the International Space Station or space shuttle. Microbiol Immunol 60:295–302. https://doi.org/10.1111/1348-0421.12375
Scher RK (1994) Onychomycosis is more than a cosmetic problem. Br J Dermatol 130:15–26
Schwarz L (1981) Histoplasmosis. Praeger Science Press, New York
Semel JD, Goldin H (1996) Association of athlete's foot with cellulitis of the lower extremities: diagnostic value of bacterial cultures of ipsilateral interdigital space samples. Clin Infect Dis 23:1162
Shalchian-Tabrizi K, Minge MA, Espelund M, Orr R, Ruden T et al (2008) Multigene phylogeny of Choanozoa and the origin of animals. PLoS One 3:e2098. https://doi.org/10.1371/journal.pone.0002098
Sharkey-Mathis, Patricia Kay, Kauffman CA, Graybill JR, Stevens DA, Hostetler JS, Cloud G, Dismukes WE (1993) NIAID Mycoses Study Group. Treatment of sporotrichosis with itraconazole. The American journal of medicine 95:279–285
Springer J, Loeffler J, Heinz W, Schloßnagel H, Lehmann M et al (2011) Pathogen-specific DNA enrichment does not increase sensitivity of PCR for diagnosis of invasive Aspergillosis in neutropenic patients. J Clin Microbiol 49:1267–1273. https://doi.org/10.1128/JCM.01679-10
Strick LB, Wald A (2006) Diagnostics for herpes simplex virus. Mol Diagn Ther 10:17–28. https://doi.org/10.1007/BF03256439
Stringer JR, Beard CB, Miller RF, Wakefield AE (2002) A new name (*Pneumocystis jirovecii*) for Pneumocystis from humans. Emerg Infect Dis 8:891–896
Taylor DL, Hollingsworth TN, McFarland JW, Lennon NJ, Nusbaum C et al (2014) A first comprehensive census of fungi in soil reveals both hyperdiversity and fine-scale niche partitioning. Ecol Monogr 84:3–20. https://doi.org/10.1890/12-1693.1
Tong KB, Lau CJ, Murtagh K, Layton AJ, Seifeldin R (2009) The economic impact of aspergillosis: analysis of hospital expenditures across patient subgroups. Int J Infect Dis 13:24–36. https://doi.org/10.1016/j.ijid.2008.02.013
Tosti A, Piraccini BM, Lorenzi S (2000) Onychomycosis caused by nondermatophytic molds: clinical features and response to treatment of 59 cases. J Am Acad Dermatol 42:217–224
Vazquez JA (1999) Options for the management of mucosal candidiasis in patients with AIDS and HIV infection. Pharmacotherapy 19:76–87
Walsh TJ, Dixon DM (1996) Spectrum of mycoses. Med Microbiol 4:91

Wheat LJ, Slama TG, Zeckel ML (1985) Histoplas-mosisinthe acquired immune deficiency syndrome. Am J Med 78:3–21

White TJ, Bruns T, Lee S, Taylor J (1990) Amplification and direct sequencing of fungal ribosomal RNA genes for phylogenetics. In: Innis MA et al (eds) PCR protocol: a guide to methods and applications. Academic Press, San Diego, pp 315–322

Wilson DA, Joyce MJ, Hall LS, Reller LB, Roberts GD et al (2005) Multicenter evaluation of a Candida albicans peptide nucleic acid fluorescent in situ hybridization probe for characterization of yeast isolates from blood cultures. J Clin Microbiol 43:2909–2912. https://doi.org/10.1128/JCM.43.6.2909-2912.2005

Woods JP (2016) Revisiting old friends: developments in understanding Histoplasma capsulatum pathogenesis. J Microbiol 54:265–276

Woolley PD, Higgins SP (1995) Comparison of clotrimazole, fluconazole and itraconazole in vaginal candidiasis. Br J Clin Pract 49:65–66

# Fungal Infections Associated with Primary and Secondary Immunodeficiencies

**Ana K. Galván-Hernández, Manuela Gómez-Gaviria, and Héctor M. Mora-Montes**

## Introduction

In recent years, fungal infections have caused more than 1.6 million deaths per year, and according to statistics, more than one billion people suffer from fungi-related diseases, which can trigger serious health problems (Brown et al. 2012; Almeida et al. 2019). Although fungi are generally associated with superficial infections in humans, some factors, such as alterations in the immune response, changes in microbiota, and predisposition to some diseases, can be the triggers of invasive mycoses (Antachopoulos 2010). The integrity of the immune system is of great importance for the defense against different infectious microorganisms, including fungi. When the host immunity is impaired, the pathogen can cause severe and often fatal medical conditions, i.e., immunodeficiency diseases.

Immunodeficiencies (ID) can be caused by quantitative or functional changes in the different mechanisms involved in either the innate or adaptive immune response (Sánchez-Ramón et al. 2019). IDs can be primary if they are of genetic origin or secondary when they are acquired. Both types are associated with infections, autoimmune disorders, inflammatory disorders, immune dysregulation, and cancer, among others (Picard et al. 2018). Thus far, the Primary Immunodeficiency Classification Committee of the International Union of Immunological Societies has identified eight groups of primary immunodeficiencies (PID) and has classified them according to the underlying immune disorder or the most predominant symptoms. Within these PIDs, the most frequent are antibody deficiencies, phagocytic function defects, and syndromes that are already well-defined (Picard et al. 2018). On the other hand, secondary immunodeficiencies (SID) are associated with

A. K. Galván-Hernández · M. Gómez-Gaviria · H. M. Mora-Montes (✉)
Departamento de Biología, División de Ciencias Naturales y Exactas,
Campus Guanajuato, Universidad de Guanajuato, Guanajuato, Mexico
e-mail: hmora@ugto.mx



A. Gomes Rodrigues, A. Gupta (eds.), *Fungal Infections*, Fungal Biology,
https://doi.org/10.1007/978-3-032-06014-3_2

systemic disorders, where hematological conditions, medications, or viral infections are the underlying disease, and generate severe or chronic long-term diseases (Chinen and Shearer 2010).

Most PIDs are identified during childhood and are often associated with recurrent infections and changes in the immune response, where autoimmunity, allergies, and autoinflammation are included (Modell et al. 2018). Thus far, more than 450 genetically defined diseases are included within the PIDs classification (Bousfiha et al. 2015), and it is estimated that they are more common than estimated, affecting approximately 1% of the population (Boyle and Buckley 2007; Bousfiha et al. 2013). PIDs are categorized according to immune system involvement (Table 1): defects in innate immunity, antibody deficiencies, combined immunodeficiencies, defects in T lymphocyte activation and function, and multisystem disorders with immunodeficiency (Picard et al. 2015).

In the case of SID, it is defined as a transient or persistent impairment of the function of immune cells or tissues, which can be caused by factors outside immunity (Tuano et al. 2021). These factors include environmental agents, drugs, and a variety of heterogeneous conditions, such as prematurity and aging, genetic syndromes, infections, chemotherapy, radiotherapy, corticosteroids, malnutrition, leukemia, enteropathies, nephropathies, UV light, and hypoxia (Table 1) (Tuano et al. 2021).

Immunodeficiencies lead to the establishment of opportunistic infections caused by microorganisms such as fungi, viruses, and bacteria. There are a large number of fungal species that are associated with human infections, and the clinical manifestation and severity of these mycoses are influenced by the host defense mechanisms and the fungal strain (Romani 2004). Among the most known genus associated with invasive mycoses in immunodeficient populations are *Candida*, *Aspergillus*, *Histoplasma*, *Cryptococcus*, *Pneumocystis*, *Paracoccidioides,* and *Mucor* (Badiee and Hashemizadeh 2014; José and Brown 2016).

Fungal diseases caused by these fungal genera have a major impact on human health. About 220,000 cases of cryptococcal meningitis occur each year worldwide, and more than 400,000 people develop *Pneumocystis*-caused pneumonia (Rajasingham et al. 2017; Rodrigues and Nosanchuk 2020). In Latin America, histoplasmosis is considered a common opportunistic infection, especially in patients with HIV (Bongomin et al. 2017; Rodrigues and Nosanchuk 2020); and candidiasis remains the most frequent cause of invasive fungal infections, with an incidence estimated at 73 cases per million inhabitants/year, surpassing invasive aspergillosis and mucormycosis (Guinea 2014). This chapter will review the incidence and clinical manifestations of the most recurrent fungal diseases in patients with different types of immunodeficiency.

**Table 1** Main characteristics of primary and secondary immunodeficiencies

| Immunodeficiency | Functional deficiencies | Causative agents of infection | References |
|---|---|---|---|
| *Defects of innate immunity* | | | |
| Chronic granulomatous disease (CGD) | Dysfunction of the NADPH enzyme complex | *Aspergillus fumigatus*<br>*Aspergillus nidulans*<br>*Paecilomyces*<br>*Fusarium*<br>*Scedosporium*<br>*Paecilomyces variotti*<br>*Paecilomyces lilacinus* | Winkelstein et al. (2000); Espinoza et al. (2015); Roos (2016) |
| CARD9 deficiency | Decreased production of inflammatory cytokines | *Candida albicans*<br>*Candida glabrata*<br>*Candida dubliniensis*<br>*Candida tropicalis*<br>*Dermatophytes*<br>*Cryptococcus neoformans*<br>*A. fumigatus* | Rieber et al. (2016); Corvilain et al. (2018); Drummond et al. (2018) |
| IL-12 deficiency | Defects in NK and T lymphocyte activation | *C. albicans*<br>*C. neoformans*<br>*Paracoccidiodes brasiliensis* | Emmanuel et al. (1999); Jirapongsananuruk et al. (2012) |
| *Combined immunodeficiencies* | | | |
| MALT1 deficiency | Deficiency in T and B cell activation | *C. albicans*<br>*C. parapsilosis*<br>*Aspergillus* spp. | Shirmast et al. (2020); Lu and Turvey (2021); Sefer et al. (2022) |
| DOCK8 deficiency | Defects in innate cell functions and cytoskeleton rearrangements | *Candida* spp.<br>*Aspergillus* spp. | Vinh (2011); Alcántara-Montiel and Vega-Torres (2016) |
| APECED | Infiltration of lymphocytes in affected organs, loss of endocrinological functions | *Candida* spp. | Perheentupa (2006) |
| Mutations in *IKBKB* | Defects in T and B lymphocyte function | *C. albicans*<br>*Pneumocystis jirovecii* | Nielsen et al. (2014); Cuvelier et al. (2019) |
| *Other primary immunodeficiencies* | | | |
| STAT3 deficiency | Defects in cell signaling and immune response | *Candida* spp. | Chandesris et al. (2012) |

(continued)

**Table 1** (continued)

| Immunodeficiency | Functional deficiencies | Causative agents of infection | References |
|---|---|---|---|
| CD18 deficiency | Leukocyte adhesion and migration defects | *Candida* spp. | Kauer et al. (2021) |
| GATA 2 deficiency | Defects in the development and function of blood cells and the immune system | *Aspergillus* spp.<br>*Cryptococcus* spp. | Iseki et al. (1994); Hsu et al. (2011) |
| INF-γ/IL-12 deficiency | Defects in the immune system's ability to fight infection by bacteria and viruses | *Coccidioides immitis*<br>*Coccidioides posadii*<br>*P. brasiliensis*<br>*Histoplasma capsulatum* | Zerbe and Holland (2005); Vinh (2011) |
| *Malnutrition and metabolic disorders* | | | |
| Malnutrition | Altered immune responses and lymphocytes, macrophages, and granulocytes | *C. albicans*<br>*C. tropicalis*<br>*Candida krusei*<br>*C. glabrata*<br>*Candida parapsilosis*<br>*Candida guilliermondii*<br>*Trichosporon* spp.<br>*Pneumocystis jirovecii*<br>*H. capsulatum*<br>*Lichtheimia ramosa* | Field et al. (2002); Paillaud et al. (2004); Schaible and Kaufmann (2007); Hanachi et al. (2018); Colman et al. (2022) |
| Diabetes mellitus | Changes in cellular immunity, involvement of polymorphonuclear cells, monocytes, and lymphocytes | *C. albicans*<br>*C. glabrata*<br>*C. tropicalis*<br>*C. parapsilosis*<br>*C. neoformans*<br>*A. fumigatus*<br>*Rhizopus* spp.<br>*Mucor* spp.<br>*Rhizomucor* spp. | Calvet and Yoshikawa (2001); Lao et al. (2020); Corzo-León et al. (2018) |
| Chronic kidney disease | Defects in immune function, impaired cellular and humoral immunity. Altered phagocytosis and T cell function | *C. albicans*<br>*C. tropicalis*<br>*C. parapsilosis*<br>*C. neoformans*<br>*Coccidioides* spp.<br>*Fusarium* spp.<br>*Aspergillus* spp. | Abbott et al. (2001); Prasad et al. (2004); Gandhi et al. (2005) |
| *Infections* | | | |

(continued)

**Table 1** (continued)

| Immunodeficiency | Functional deficiencies | Causative agents of infection | References |
|---|---|---|---|
| HIV | Immune system dysfunction, decreased CD4 cell count | *P. jirovecii*<br>*C. neoformans*<br>*H. capsulatum*<br>*Talaromyces*<br>*C. albicans*<br>*C. tropicalis*<br>*C. krusei*<br>*C. glabrata*<br>*A. fumigatus*<br>*Sporothrix schenckii*<br>*Penicillium marneffei* | Le et al. (2010); Freitas et al. (2014); Jain et al. (2014); Kaur et al. (2016); Limper et al. (2017) |
| *Temporary immunodeficiency* | | | |
| COVID-19 | Increased proinflammatory markers (IL-1, IL-6, and TNFα), lower CD4 and CD8 cell counts, and decreased CD4 interferon-gamma expression | *Rhizopus* spp.<br>*C. albicans*<br>*C. auris*<br>*C. tropicalis*<br>*Aspergillus* spp. | Chowdhary et al. (2020); Chakrabarti (2021); Yusuf et al. (2021) |
| *Trauma, burns, and major surgery* | | | |
| Organ transplantation | Long hospital stays, allograft damage, high mortality rates | *H. capsulatum*<br>*Blastomyces dermatitis*<br>*C. immitis*<br>*C. neoformans*<br>*C. immitis*<br>*Scedosporium apiospermun*<br>*C. albicans*<br>*C. parapsilosis*<br>*C. glabrata*<br>*C. krusei*<br>*C. guilliermondii*<br>*A. fumigatus*<br>*A. flavus*<br>*A. niger*<br>*A. terreus*<br>*C. neoformans*<br>*Cryptococcus gattii* | Walsh et al. (2008); Pappas et al. (2009); Baddley et al. (2010); Harris et al. (2011); Shoham and Marr (2012) |
| *Immunosuppressive medications* | | | |
| Chemotherapy, anti-inflammatory and immunomodulatory drugs | Changes in the immune system, defects in the phagocytic function of monocytes and neutrophils. Alterations of antigen presentation, and in the secretion of proinflammatory cytokines. | *C. albicans*<br>*C. glabrata*<br>*C. krusei*<br>*C. neoformans*<br>*A. fumigatus*<br>*A. flavus* | Steinbach et al. (2012); Sun et al. (2015); Schmidt et al. (2019) |

# Primary Immunodeficiencies (PID)

Some of the most common PIDs associated with fungal infections are chronic granulomatous disease (CGD), CARD9 and IL-12 deficiency, and combined immunodeficiencies, among others. The fungal infections most frequently associated with these IDs are those caused by *Aspergillus*, *Candida,* and *Cryptococcus* (Antachopoulos et al. 2007). The following sections will provide information on the most common fungal infections that occur in patients with PIDs. The prevalence of the different fungal pathogens in PIDs is shown in Fig. 1.

## *Defects of Innate Immunity*

### Chronic Granulomatous Disease

Chronic granulomatous disease (CGD) is an inherited immunodeficiency due to the specific disruption of the X-linked gene *gp91*$^{phox}$, which is characterized by dysfunction of the nicotinamide adenine dinucleotide phosphate oxidase (NADPH) enzyme complex (Espinoza et al. 2015). As a consequence, phagocytic cells cannot

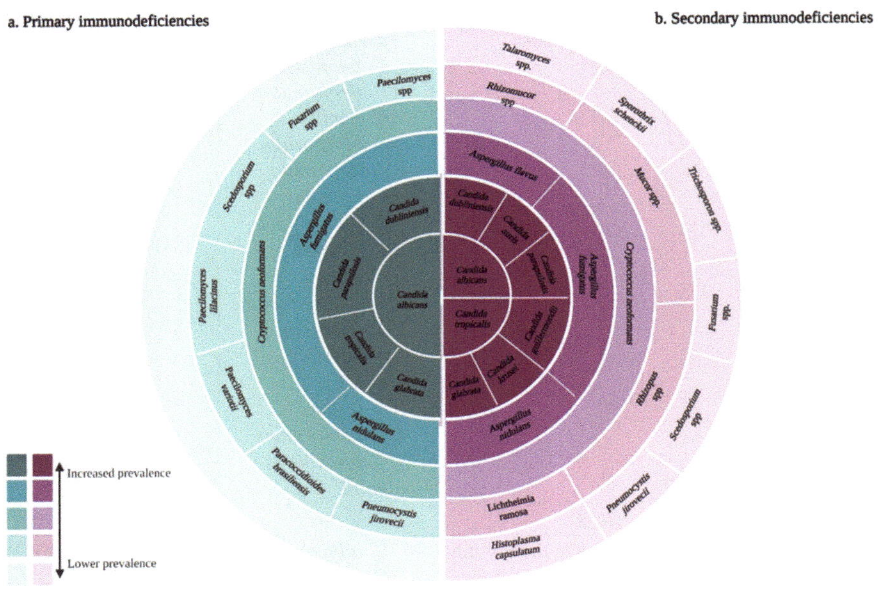


**Fig. 1** Pathogenic fungi associated with primary and secondary immunodeficiencies. (**a**) Prevalence of different fungal species in primary immunodeficiencies. (**b**) Prevalence of different fungal species in secondary immunodeficiencies. The colored bars indicate the most prevalent and least prevalent species in the different immunodeficiencies

generate nicotinamide adenine dinucleotide phosphate, which is important for the intracellular death of the phagocytosed microorganisms, leading to recurrent infections and the formation of inflammatory granulomas in the affected tissues (Espinoza et al. 2015). Invasive fungal infections, especially those caused by species such as *Aspergillus* and *Candida*, are the leading cause of mortality in patients with CGD. In particular, 35% of infection-related deaths in these patients are due to *Aspergillus* species such as *Aspergillus fumigatus* and *Aspergillus nidulans*, which can cause pneumonia, lung abscesses, and brain abscesses. In addition, cases of osteomyelitis or contiguous invasion of bone have been reported, typically associated with non-*fumigatus Aspergillus* pneumonia. Infections caused by *A. nidulans* can reach the ribs or vertebral bodies of patients with this type of ID (Winkelstein et al. 2000; Roos 2016). The diagnosis to detect these pathogens is based on microscopic examination and culture of blood and respiratory samples and biopsies; in addition, computed tomography or magnetic resonance imaging is useful to detect lung lesions (Denning et al. 2013). Serological tests can also help in the diagnosis of invasive aspergillosis by detecting specific antibodies, such as those against galactomannan (Hoenigl et al. 2014).

The use of antifungal drugs, such as itraconazole, voriconazole, and posaconazole, has been shown to decrease mortality rates in infections caused by *Aspergillus* in patients with CGD (Espinoza et al. 2015). However, treatment of fungal infections in these types of conditions can pose a challenge due to the prolonged duration of treatment which can be from 3 to 6 months (Roos 2016). Immunomodulatory therapy, such as interferon-gamma, can improve immune function and decrease the frequency of *Aspergillus* infections in CGD patients (Ahlin et al. 1999). Rarely, *Paecilomyces*, *Fusarium*, *Scedosporium*, *Paecilomyces variotti*, and *Paecilomyces lilacinus* infections occur, which can cause severe and even fatal pneumonia (Williamson et al. 1992; Yu et al. 2021).

## CARD9 Deficiency

CARD9 protein is an intracellular adaptor protein that belongs to the CARD family of proteins. This protein functions as a scaffolding protein that transmits signals from Toll-like receptors (TLRs) and C-type lectin receptors (CLRs) to mitogen-activated protein kinase (MAPK) and nuclear transcription factor (NF-κB). Thus, CARD9 is a link between innate and adaptive immunity (Wang et al. 2020). Pathogenic fungi are primarily recognized by CLRs and TLRs, many of which signaling via CARD9 to initiate defense against fungal invasion (Drummond et al. 2018). Deficiency of the gene encoding CARD9 is associated with increased susceptibility to fungal infections, including those caused by *C. albicans*, dermatophytes, *C. neoformans*, and *A. fumigatus*. This deficiency can lead to low production of inflammatory cytokines and chemokines, which can lead to an inadequate immune response during fungal infection and a lack of regulation of inflammation in affected tissues (Lanternier et al. 2013).

Autosomal recessive CARD9 deficiency is associated with extrapulmonary aspergillosis due to impaired neutrophil accumulation at sites of infection (Rieber et al. 2016). Fungal infections in patients with CARD9 deficiency are commonly caused by *Candida* species, where hyphae play a key role in tissue invasion (Corvilain et al. 2018). In addition to *C. albicans*, other species such as *C. glabrata*, *C. dubliniensis,* and *C. tropicalis* have also been reported to cause fungal infections in these patients (Corvilain et al. 2018). CARD9 deficiency manifests clinically in various forms of candidiasis, with *C. albicans* being the most common species affecting mainly the oral cavity, vaginal mucosa, gastrointestinal tract, and skin (Glocker et al. 2009). A study conducted with CARD9-deficient neutrophils showed that they are unable to effectively clear *Candida* species, which can lead to implantation and the spread of infection (Drewniak et al. 2013). The diagnosis of candidiasis in patients with CARD9 deficiency is performed through the culture and microscopic analysis of samples, and serological tests using anti-*Candida* antibodies. In these cases, it is very important to carry out an accurate identification of the species involved, since its treatment will depend on this (Drummond et al. 2018). Antifungals, such as itraconazole, voriconazole, fluconazole, and amphotericin B have been used in these patients. However, it has been reported that some patients may have increased susceptibility to some antifungals, such as amphotericin B, which may require lower doses or monitoring for side effects. In some cases, the use of immunomodulatory therapies is recommended to improve immune system functions (Glocker et al. 2009; Lanternier et al. 2013; Corvilain et al. 2018).

In addition to causing recurrent infections in the lungs of patients with CARD9 deficiency, *A. fumigatus,* and *C. neoformans* can affect the central nervous system (Rieber et al. 2016; Wang et al. 2020). A study in African families with CARD9 deficiency and suffering from recurrent fungal infections showed that clinical signs of deep dermatophytosis started in childhood with recurrent onychomycosis and ringworm, chronically evolving into an invasive disease, with 75% survival of patients (Lanternier et al. 2013). Likewise, dermatophytosis can also present as Majocchi granuloma and deep dermatophytosis with nonlocalized dermal invasion, which can extend beyond the perifollicular area (Ilkit et al. 2012; Lanternier et al. 2013). Regarding the diagnosis of these mycoses, methods similar to those mentioned above for candidiasis can be employed, where clinical evaluation, microscopic examination, and culture of clinical specimens are included. Treatment may include specific antifungals for each type of mycosis, such as voriconazole and amphotericin B for *Aspergillus* and *Cryptococcus*, respectively.

## IL-12 Deficiency

IL-12 is produced by antigen-presenting cells and phagocytes in response to pathogens and activates NK and T lymphocytes by inducing INF-γ production and cytolytic activity, making it a proinflammatory cytokine. In addition, it induces $T_H1$ lymphocyte differentiation, phagocytic cell activation, generation of cytotoxic T cells, and production of opsonizing antibodies (Méndez-Samperio et al. 1998).

The association of IL-12 deficiency and *C. albicans* infections has been extensively studied. A study that analyzed 128 patients with IL-12Rβ1 deficiency found that 25% of them (32 patients) developed candidiasis. Among the cases studied, 76 episodes of candidiasis were reported, of which 53 corresponded to recurrent infections and 23 to isolated infections. The oropharyngeal cavity was the most frequent site of infection, occurring in 78% of cases, while the remaining sites were the esophagus, vulva/vagina, and skin (Ouederni et al. 2014). The treatment involved the use of fluconazole, voriconazole, itraconazole, amphotericin B, and echinocandins. Higher doses of antifungals and longer treatment were required compared to patients without IL-12 deficiency (Ouederni et al. 2014). This is related to the fact that IL-12 deficiency implies a decreased ability of the immune system to fight *C. albicans* infection, which increases the likelihood of persistent or recurrent infection (Ouederni et al. 2014; Tissot et al. 2017). Therefore, treatment of candidiasis in patients with IL-12 deficiency may be more difficult and prolonged.

There are also cases of *C. neoformans* infection causing chronic osteomyelitis. This disease is rare and occurs in only 5–10% of people with disseminated *Cryptococcus* infection, and few cases have been described in IL-12Rβ1 deficiency (Jirapongsananuruk et al. 2012). In addition, cases of *Paracoccidioides brasiliensis* infections have been reported in patients with IL-12/IL-23Rβ1 deficiency, who presented symptoms such as persistent fever, abdominal pain, disseminated lymphadenopathy and hepatosplenomegaly (Moraes-Vasconcelos et al. 2005). For the treatment of these mycoses, the use of amphotericin B and flucytosine has been reported, followed by maintenance therapy with fluconazole for *C. neoformans*. For the case of *P. brasiliensis* infections, the treatment includes the administration of itraconazole or posaconazole for a prolonged period (Ouederni et al. 2014; Mendes et al. 2017).

## *Combined Immunodeficiencies*

### MALT1 Deficiency

MALT1 is a cytoplasmic protease that binds to BCL10, which is in proenzyme form. Together, these proteins form a complex known as CBM (CARD-BCL10-MALT1), which is recruited for NFκB activation, as well as activation of B and T lymphocytes in response to stimulation of antigen receptors (such as TCR or BCR), ITAMs, or GPCRs (Rosebeck et al. 2011). Incorporation of CARD proteins into BCL10 leads to oligomerization of BCL10, which forms a filamentous structure that serves as a scaffold for various signaling complexes, including IKK-dependent ones, which are crucial for NFκB activation. As a result, MALT1 deficiency leads to impaired T and B cell activation (Juilland and Thome 2018).

MALT1 deficiency is characterized by immunological features, such as a normal lymphocyte count, normal or low immunoglobulin levels, low titers of antigen-specific antibodies, and a defective T-cell proliferative response to mitogenic stimuli. In addition, cases with increased T-cell numbers and impaired B-cell maturation have been reported (Lu et al. 2018). This deficiency may also lead to a reduction in regulatory (Treg) and Th17 T cells (Lu et al. 2018).

The development of bacterial, viral, and fungal infectious diseases has been described in patients with MALT1 deficiency. Although fungal organisms are not usually detected, the presence of *C. albicans* has been reported in individuals with this deficiency. This species causes skin infection in conjunction with *Staphylococcus aureus* and cytomegalovirus in 64% of cases and has been isolated in 27% of cases of oral cavity infection (Lu and Turvey 2021). A cohort study of 19 patients with this deficiency found a 78% prevalence of *C. albicans*, although *Candida parapsilosis*, which caused septic arthritis, was also found (Sefer et al. 2022). Likewise, it has been found that *Aspergillus* can cause skin infections (Sefer et al. 2022). Cases of *Pneumocystis* pneumonia have been reported in infants under one year old, with agammaglobulinemia, due to the inability to activate B cells (Sonoda et al. 2021). The diagnosis of candidiasis and aspergillosis in patients with MALT1 deficiency is performed by the detection of the causative agent in clinical specimens, such as cultures and serological samples. For *Pneumocystis*, silver staining microscopy and nucleic acid amplification tests can be used for detection (Jongco et al. 2014; McKinnon et al. 2014; Lu and Turvey 2021). The treatment of these mycoses for the case of candidiasis and aspergillosis in patients with MALT1 deficiency is based on the administration of fluconazole, voriconazole, itraconazole, and amphotericin B, administered orally or intravenously, depending on the severity of the infection (Lu and Turvey 2021). The treatment may extend from weeks to months. It is important to note that in patients with MALT1 deficiency, treatment of these infections may be more complex due to susceptibility to recurrent infections and possible resistance to antifungal drugs.

## DOCK8 Deficiency

Combined immunodeficiencies are characterized by the inability to generate B and T lymphocytes, as well as natural killer (NK) lymphocytes. In these patients, it is common to find disseminated invasive or mucous membrane infections, with *Candida* being the most frequently implicated fungal etiological agent (Vinh 2011). DOCK8 is a protein member of the DOCK180 family of guanine nucleotide exchange factors that interact with Rho GTPases and regulate cytoskeleton rearrangements necessary for cell structure, migration, adhesion, and other functions in innate cells, mainly lymphocytes (Vinh 2011). DOCK8 deficiency is associated with autosomal recessive hyper-IgE syndrome, which is characterized by the presence of asthma and severe allergies, as well as the frequent occurrence of mucocutaneous infections (Zhang et al. 2009).

Patients with DOCK8 antibody deficiency are at increased risk for fungal infections, particularly *Candida* and *Aspergillus* infections, as well as those caused by *Cryptococcus*, *Trichosporon*, *Rhizopus*, and *Penicillium* (Winkelstein et al. 2003; Zhang et al. 2009). A systematic review reported that the most common fungal infection in patients with DOCK8 deficiency is invasive candidiasis (Su et al. 2011). In addition, these patients have been described to be at increased risk of developing invasive pulmonary aspergillosis and cryptococcosis (Alcántara-Montiel and Vega-Torres 2016). However, more information is needed to know in detail how these mycoses affect patients with deficiencies in this protein. The diagnosis of these mycoses is performed conventionally by culturing on Sabouraud agar for *Candida* and *Aspergillus*, or CHROMagar, which is more specific for *Candida*. In the case of aspergillosis, tissue or body fluid cultures can be performed, such as bronchoalveolar lavage or aspirate of lung lesions (Zhang et al. 2009). The treatment of candidiasis and aspergillosis in patients with DOCK8 deficiency is based on the use of fluconazole, voriconazole, itraconazole, and amphotericin B, administered orally or intravenously, depending on the severity of the infection. The treatment may be prolonged and depends on the patient's response. Hematopoietic stem cell transplantation may be considered a therapeutic option in patients with recurrent fungal infections resistant to conventional treatments (Lanternier et al. 2013; Lionakis 2019).

## Autoimmune Polyendocrinopathy Candidiasis Ectodermal Dystrophy (APECED)

Autoimmune polyendocrinopathy-candidiasis-ectodermal dystrophy (APECED) is a genetic disease characterized by the presence of hypoparathyroidism, primary adrenocortical insufficiency, and chronic mucocutaneous candidiasis (Ahonen 1985). In this pathology, there is the production of tissue-specific autoantibodies, which are responsible for the infiltration of lymphocytes in the affected organs, causing long-term damage and loss of their endocrine functions (Heino et al. 2001). This disease is caused by a mutation in the AIRE gene, which is predominantly expressed in the thymus, pancreas, and adrenal cortex. This mutation consists of a C → T transition at nucleotide 889 in exon 6, leading to the change of an Arg to a premature stop codon, resulting in a truncated protein (Björses et al. 1998). APECED is associated with a deficiency of T lymphocytes and autoantibodies against IL-17 and IL-22, which are critical for mucosal antifungal immunity (Conti et al. 2009). As a result, increased susceptibility to chronic mucocutaneous candidiasis occurs.

Chronic mucocutaneous candidiasis is the first disease to appear in the triad of the main components of APECED (Perheentupa 2006). This infection manifests with the onset of oral candidiasis, which can cause ulcers in the mouth. In the hyperplastic form, the tongue and mucous membranes show hyperkeratosis and white or gray plaques, while in the atrophic form, it may appear as a generalized red plaque with the presence of nodules that may be potentially carcinogenic (Husebye et al. 2009). In addition, this infection can extend to the esophagus and intestines, and although intestinal candidiasis can occur in the absence of oral disease, it can also manifest with skin infections, mainly on the hands and nails. In some cases, the clinical manifestations also involve vaginal candidiasis (Kisand and Peterson 2015). Considering that species of the *Candida* genus are those commonly associated with this ID, the diagnosis is usually established through family and personal history, serological tests to detect tissue-specific autoantibodies, as well as cultures of clinical specimens, such as saliva (Puel et al. 2010; Capalbo et al. 2012). In terms of treatment, the use of antifungals is the main strategy. Patients may receive azole therapy, such as fluconazole, itraconazole, or voriconazole, and in some cases, the use of amphotericin B may be necessary. Patients with APECED may require long-term treatment and preventive therapy against recurrent candidiasis (Capalbo et al. 2012).

## Mutations in *IKBKB*

Mutations that alter the NF-κB signaling pathway can cause immunological diseases. Several studies have shown that mutations in the *IKBKB* gene, which encodes the *IKK2* protein, cause defects in T and B lymphocyte function, leading to severe combined immunodeficiencies (Pannicke et al. 2013; Mousallem et al. 2014; Cardinez et al. 2018).

*IKK2* deficiency has a direct impact on the signaling of surface receptors of T, B, and NK cells, significantly reducing their activation. Patients with mutations in the *IKBKB* gene are particularly susceptible to developing oral, oropharyngeal, and genitourinary candidiasis, accompanied by hypogammaglobulinemia or agammaglobulinemia (Pannicke et al. 2013).

In neonates and infants, deficiency in NF-κB activation can lead to the development of *Pneumocystis jirovecii* pneumonia (Nielsen et al. 2014). In addition, in children under one-year-old with hypogammaglobulinemia or other deficiencies in NF-κB activation, *C. albicans* infection is common and can infect the oropharynx, perineum, and other areas of the body (Mousallem et al. 2014). In a study of 14 patients, 11 had *C. albicans* infections, and signs of disseminated infection were found in 3 of them, in blood, urine, and cerebrospinal fluids, along with infections by other pathogens (Cuvelier et al. 2019). The diagnosis of candidiasis and pneumonia in patients with mutations in the *IKBKB* gene is usually performed by identification of clinical symptoms, physical examination, and results of laboratory tests, such as cultures of clinical specimens or serological tests. In some cases, genetic testing may also be performed to identify mutations in *IKBKB* (Cuvelier et al. 2019; Lionakis 2019).

Treatment for these patients usually involves the use of antifungals and antibiotics. Antifungals, such as fluconazole or voriconazole, can be administered orally or intravenously to treat candidiasis in different parts of the body, such as the mouth, throat, or genitals (Jung et al. 2020). *P. jirovecii* pneumonia is often treated with trimethoprim-sulfamethoxazole, although other drugs may be used in severe cases. In addition to the treatment of the infection, gamma globulins may be administered to replace antibody deficiency in patients with hypogammaglobulinemia. Supportive measures, such as supplemental oxygen or mechanical ventilation, may also be required to treat severe pneumonia (Pannicke et al. 2013; Picard et al. 2015).

## *Other Primary Immunodeficiencies*

*Candida* spp. is a common microorganism found on the skin and digestive tract of healthy individuals. However, in people with alterations in their immune system, it can trigger mucocutaneous infections ranging from acute to chronic (Gómez-Gaviria et al. 2023). T-lymphocyte deficiency is the main cause of chronic mucocutaneous candidiasis in people with combined immunodeficiencies (Puel et al. 2012). In the case of hyper-IgE syndrome and STAT3 AD deficiency, 85% of patients develop chronic mucocutaneous candidiasis, and 64% show oral candidiasis in newborns (Chandesris et al. 2012). AR type 1 leukocyte adhesion deficiency with CD18 deficiency can trigger invasive candidiasis due to the inability of phagocytic cells to adhere to endothelial cells and migrate to the infection site of infection (Kauer et al. 2021).

Individuals with AD GATA2 deficiency have an increased incidence of invasive aspergillosis and are more susceptible to fungal infections due to nonsense mutations affecting the zinc-finger2 domain of GATA2 (Hsu et al. 2011). This syndrome is characterized by B and NK lymphocytopenia, as well as T lymphocyte involvement, and by the presence of abnormal granules in neutrophils with abnormal surface antigen expression (Vinh et al. 2010). In addition, cryptococcosis has been described in individuals with idiopathic CD4 lymphopenia and AD GATA2 deficiency, as well as in some individuals with X-linked CD40L deficiency (Iseki et al. 1994).

Disseminated fungal infections, such as those caused by *Coccidioides immitis, Coccidioides posadasii, Paracoccidioides brasiliensis,* and *Histoplasma capsulatum,* can occur in patients with deficiencies in the INF-$\gamma$/IL-12 pathway due to mutations affecting this pathway (Vinh 2011).

Finally, people with primary immunodeficiencies have an increased risk of contracting fungal infections, so it is important to take preventive measures to reduce that risk. These include good personal hygiene, such as regular hand washing, avoiding sharing personal items, and keeping skin and mucosal areas clean. In addition, patients must receive appropriate vaccinations to prevent viral infections that can further weaken the immune system and increase the risk of fungal infections. In some cases, prophylactic antifungal therapies can be used to reduce the risk of fungal infections in patients with primary immunodeficiencies. Prophylactic antifungals are administered at low doses to prevent colonization and fungal infection in patients at high risk, such as those with prolonged neutropenia, bone marrow transplantation, or immunosuppressive treatments. However, it is important to keep in mind that prolonged use of prophylactic therapies may increase the risk of antifungal resistance and side effects. Therefore, it is essential to carefully evaluate the risk and benefits of prophylactic therapy in each case (Bonilla et al. 2015; Picard et al. 2015).

## Secondary Immunodeficiencies (SID)

Fungal infections can be a significant problem in patients with SID, as these conditions can also weaken the immune system and increase the infection risk. SID can be caused by a variety of factors, including chronic diseases, such as diabetes and HIV infection/AIDS, treatment with chemotherapy or radiation therapy for cancer, organ or tissue transplantation, and immunosuppressive drugs used in the treatment of autoimmune diseases (Axelrod and Adams 2019; Cai and Sereti 2021; Tuano et al. 2021). In these patients, fungal infections may be more difficult to treat and may increase the risk of serious complications. These patients must be carefully monitored and receive prompt and effective treatment for fungal infections to prevent serious complications and reduce the mortality risk (Sánchez-Ramón et al. 2019). The immune defects observed in this type of SID show different clinical manifestations, and the prognosis depends on the patient's immune status (Sánchez-Ramón et al. 2019). The following is a compilation of information on the most common fungal infections that have a higher frequency of occurrence in people with some form of SID.

The prevalence of the different fungal pathogens present in SID is shown in Fig. 1.

### *Malnutrition and Metabolic Disorders*

#### Malnutrition

Malnutrition has been recognized as one of the most frequent causes of ID and is known to affect many communities worldwide, which have restricted access to good nutrition (Rahman and Adjeroh 2015; Bourke et al. 2016). This is responsible for 54% of the 10.8 million deaths per year in children under 5 years of age (Schaible and Kaufmann 2007). Different works have shown that protein, calorie, and micronutrient deficiencies can alter immune responses and represent a key factor in susceptibility to different infections (Beisel 1996). In patients with advanced malnutrition, both acquired immunity and innate host defense mechanisms are affected. The functions of lymphocytes, macrophages, and granulocytes decrease, making these patients more susceptible to attack by opportunistic pathogens and their subsequent elimination (Schaible and Kaufmann 2007). Malnutrition during the early stages of life affects thymus development, resulting in defects in immunity due to a long-term reduction in lymphocyte counts (CD4 and CD8) (Savino 2002). There is an increase in CD4/CD8 double negative T cells, and therefore, immature T cells appear in the periphery. Immune defense in the epithelial barrier of a malnourished host is compromised, affecting the intestinal mucosa, with reduced lymphocyte counts and reduced immunoglobulin A (IgA) secretion (Schaible and Kaufmann 2007). These events cause opportunistic pathogens, such as *Candida* spp. to affect and invade healthy tissues. Malnutrition has been reported to increase the risk of oral colonization by *Candida*, and different species can affect its mucosa, like *C. albicans*, *Candida glabrata*, *Candida krusei*, *Candida parapsilosis*, and *Candida tropicalis* (Jabra-Rizk et al. 2001). These findings are of great importance since the colonization of the oral mucosa of malnourished children by non-*albicans* species increases the likelihood of patients developing oral candidiasis. This condition can become difficult to treat, due to the increased resistance of these species to different antifungal agents (Jabra-Rizk et al. 2001).

Studies carried out in a Pakistani community with children aged 0 to 15 months with malnutrition and diarrhea found that *Candida* spp. was the most frequent microorganism isolated from stool samples, being *C. tropicalis* the most prevalent species (Klingspor et al. 1993). In a study conducted on older adults, the prevalence of oral candidiasis and its association with malnutrition was evaluated (Paillaud et al. 2004). On performing complete oral examinations to determine the microorganisms present, *Candida* species were found to be the most prevalent. In those patients with fungal disease, *C. albicans* was the most common species and was isolated from 66.6% of the samples. In six patients (16.6%), different species were isolated, the most frequent combination was *C. albicans* and *C. glabrata* (13.8%), followed by *C. albicans* and *C. tropicalis* (2.7%). In two patients (5.6%) *Candida guilliermondii* and *C. tropicalis* species were also isolated (Paillaud et al. 2004). It was further determined that oral candidiasis was common in most of these patients, and as observed in other studies, this infection could be related to malnutrition. Oral candidiasis causes mucosal lesions, which impact energy intake, causing the patient's nutritional status to worsen (Paillaud et al. 2004). Most studies affirm that in malnourished patients, *C. albicans*, *C. tropicalis,* and *C. parapsilosis* are the most frequently isolated species. The diagnosis of the different species mentioned here will depend on the type and location of the infection (Singh and Mullin 2017). In patients with malnutrition, *Candida* infections are often more difficult to diagnose due to decreased immune response. However, the diagnosis can be based on clinical history, physical examination, and laboratory tests. The most commonly used methods include culture of the infected sample, microscopic examination of the sample, and serological tests to detect antibodies against *Candida* spp. (Paillaud et al. 2004; Singh and Mullin 2017). On the other hand, treatment of this mycosis in patients with this SID should address both fungal infection and malnutrition. Treatments include the use of antifungals such as fluconazole, itraconazole, and amphotericin B. In most cases, treatment with fluconazole has proven to be effective; however, this is a long treatment scheme that can last up to three months. This may have repercussions on the health and integrity of some organs, such as the kidneys (Paillaud et al. 2004, Singh and Mullin 2017). It is important that while the fungus is being treated, malnutrition is also corrected through nutritional supplementation.

Although *Candida* species are commonly isolated from patients with malnutrition, other fungal genera also represent a health hazard in patients with these IDs. Among these, we can find *Pneumocystis* spp., *Aspergillus* spp., and *Histoplasma* spp. (Morton et al. 1980; López et al. 2016; Hanachi et al. 2018). In a patient suffering from anorexia nervosa, the opportunistic fungus *P. jirovecii* was reported to cause pneumonia and bronchitis (Hanachi et al. 2018). This patient also showed bicytopenia with a decrease in leukocytes, lymphocytes, and $CD4^+$ cells (Hanachi et al. 2018). Another fungal disease associated with this SID is invasive pulmonary aspergillosis, several reports indicate that a depression of the immune system brings about the development of this disease. One case report showed that an adolescent girl with malnutrition developed a pulmonary abscess, which was treated with amphotericin B and subsequent surgical removal (Morton et al. 1980). The infection caused by *H. capsulatum* is an important mycosis in the American continent, and in children with immune problems, the clinical manifestations of the disease are more severe (López et al. 2016). The most important risk factor is malnutrition. A retrospective study of pediatric patients who had been diagnosed with histoplasmosis showed that most cases were related to this SID (López et al. 2016). The most common clinical form in malnutrition in infant patients with histoplasmosis is disseminated disease, which affects the central nervous system and causes some neurological manifestations, such as meningeal irritation and headache (López et al. 2016). Mucormycosis has also been reported as an infection that can occur in patients who have compromised cellular immunity. A patient with acute renal hepatic failure and caloric malnutrition showed mucormycosis associated with *Lichtheimia ramosa* (Colman et al. 2022). Although several fungal genera can affect patients with malnutrition, the severity of the caused infection will depend on the patient's condition, especially the immune fitness.

## Diabetes Mellitus

Diabetes mellitus is known as a group of metabolic disorders characterized by elevated blood glucose levels. This disease is becoming one of the greatest emerging threats to public health in the twenty-first century (Rodrigues et al. 2019). Several immune alterations have been described in people with diabetes, among these, are changes in cellular immunity, where polymorphonuclear cells, monocytes, and lymphocytes are affected (Calvet and Yoshikawa 2001). Serum glucose concentrations are higher than in healthy people, in addition, these patients present a continuous loss of insulin-producing β-cells in the pancreas, which causes changes in host immune responses against different pathogens (Rodrigues et al. 2019).

People with this disease are more susceptible to acquiring different types of infections. Invasive fungal diseases (IFD) are known to be life-threatening infections, and different species of pathogenic fungi are involved. Invasive candidiasis has a mortality rate found to be around 10–15% in patients with diabetes mellitus, and invasive pulmonary aspergillosis has a mortality rate of 42–64% in those patients who are more severely ill (Bassetti et al. 2017; Saud et al. 2020). Moreover, it has been reported that when hyperglycemia is not controlled it can contribute to the poor prognosis of type 2 diabetic patients with cryptococcosis infection (Li et al. 2017).

Retrospective studies in China have investigated the clinical characteristics and factors associated with IFD in adults with type 2 diabetes. In a study that included a total of 30,984 patients with this type of diabetes, 122 were found to have IFD, indicating that the prevalence of these infections reached 0.4% (Lao et al. 2020). The most common causative agents were species of the genera *Candida*, *Aspergillus*, and *Cryptococcus*. *C. albicans* mainly affected the urinary tract and its frequency was 38.7%, followed by infections caused by *C. neoformans*, which mainly affected the lungs, causing pneumonia, and its prevalence was 64%. Mixed infections with *C. albicans*, *C. neoformans,* and *A. fumigatus* occurred in 15 patients, with *C. albicans* being the predominant species. The mortality rate in patients with IFD reached 23.3% (Lao et al. 2020). Other comorbidities were associated with IFD in these patients. Among these were anemia, hypoalbuminemia, elevated serum creatinine, and high levels of glycosylated hemoglobin (Lao et al. 2020). For the diagnosis of the different species involved, samples of sputum, ascitic fluid, urine, bronchoalveolar lavage fluid, blood, cerebrospinal fluid, and bile, among others, were obtained. In addition, biopsies were performed on 34 patients. Within the *Candida* genus, the species with the highest occurrence was *C. albicans*; however, *C. glabrata*, *C. tropicalis,* and *C. parapsilosis* were also found (Lao et al. 2020). Of the 122 patients reported with IFD, 108 received antifungal treatment; some of the antifungals used for therapy were itraconazole, flucytosine, fluconazole, and voriconazole. However, some strains of *C. albicans*, *C. glabrata,* and *C. tropicalis* were found to show resistance to some of these antifungals (Lao et al. 2020). Other studies have reported that there is a higher presence of intestinal colonization by *C. albicans* in diabetic patients (Gürsoy et al. 2018). It is also reported that *Candida* spp. ranks second among the pathogens isolated from diabetic patients, mainly affecting the urinary tract. *C. albicans* can also colonize the intestinal tract, and in patients with type I diabetes, colony-forming units of this pathogen are 2.7 times more than in healthy people (Akpan and Morgan 2002).

Other fungal species found within the class Zygomycetes also tend to occur in patients with diabetes. In Mexico, it has been reported that 72% of patients presenting with mucormycosis had diabetes mellitus; however, these figures change considering the geographic region (Corzo-León et al. 2018). In Central Europe and Asia, the figure is reduced to 17% (Rueping et al. 2010). The clinical presentations of mucormycosis and entomophthoromycosis in patients with diabetes are mostly sinusitis, followed by skin and lung infections. These infections can show complications, such as cavernous sinus thrombosis, disseminated infection, periorbital destruction, palatal ulcers, and osteomyelitis, and in more severe cases, can lead to death (Corzo-León et al. 2018). Procedures to diagnose mucormycosis include histopathology, direct cytology/smear, and cultures, and the most frequently diagnosed genera are *Rhizopus* spp., *Mucor* spp., and *Rhizomucor* spp. The most frequent treatment option is amphotericin B, and for more severe cases, surgery is used in addition to antifungal management. The latter treatment ensures a lower mortality rate compared to those patients who only received antifungal treatment (Corzo-León et al. 2018). Isavuconazole is considered another option for the primary treatment of mucormycosis; moreover, for patients with underlying diabetic nephropathy, isavuconazole is recommended to avoid damage to renal function (Desai et al. 2016). It is known that mucormycosis is more frequent in individuals with uncontrolled diabetes; thus, glycemic control is an important factor in the management of this IFD and its prevention. However, less than 40% of the population with diabetes has good glycemic control, which raises further concern as the mortality rate of these patients may increase (Brath et al. 2016).

## Chronic Kidney Disease (Uremia)

IFDs are a challenge in the management of immunocompromised patients and patients with renal disease (Jawale et al. 2020). Uremia can impair immune function and create a favorable environment for the growth of pathogenic fungi, which increases the risk of infections. In addition, patients with uremia often receive immunosuppressive and antimicrobial therapies, which increases the risk of fungal infections.

Immunological defects in people with kidney disease are related to metabolic and nutritional defects, leading to impaired cellular and humoral immunity, and the degree of impairment depends on the uremic state (Jawale et al. 2020). These defects include lymphopenia, impaired phagocytosis, and T-cell function (Jawale et al. 2020). The *Candida* genus accounts for 8 to 15% of all fungal infections acquired in patients with renal disease, especially in those patients with dialysis dependence (Sung et al. 2001). In the United States, candidiasis was reported to be the dominant infection in dialysis patients, and its frequency of occurrence was 79% (Abbott et al. 2001). In addition, other infections such as cryptococcosis and coccidioidomycosis also occur in these patients; however, the frequency is lower (Abbott et al. 2001).

Patients undergoing peritoneal dialysis may develop complications due to the appearance of fungal peritonitis, which is usually more deadly than bacterial peritonitis. The fungal species most associated with this infection are *C. albicans*, *C. tropicalis,* and *C. parapsilosis* (Nagappan et al. 1992). Other fungi, such as *Fusarium* and *Aspergillus* can contaminate and block the small orifices of peritoneal dialysis catheters (Prasad et al. 2004).

The diagnosis of these fungal infections must be immediate because this can have severe repercussions on the patient's life expectancy. One of the strategies used to determine fungal infections is to perform a peritoneal white blood cell count, and microscopic analysis is also often performed to determine the presence of yeast or hyphae (Prasad et al. 2004). Once fungal peritonitis is diagnosed, the next step is to wash the peritoneum until the returning fluid is clear, this method serves to reduce the fungal load. It is also advisable to remove the catheter, as this can be a source of infection for the patient. Systemic antifungal treatment may vary depending on the type of infection and the immune status of the patient; usually, intravenous amphotericin B, or fluconazole is used (Li et al. 2017). It is also recommended to make a combination of drugs, such as fluconazole at a concentration of 200 mg/day with flucytosine with doses of 2 g per day, for 7 days. Doses can vary, fluconazole can be increased to 400 mg/day, and if a positive response is not observed in the first 4 days, this can be replaced by amphotericin B at doses of 0.3–0.5 mg/kg. Treatment lasts for approximately 1 to 2 months (Li et al. 2017).

## *Infections*

### Human Immunodeficiency Virus (HIV)

Invasive fungal infections are one of the causes of HIV-related mortality worldwide. These infections commonly occur in individuals with advanced HIV disease; therefore, antiretroviral therapy has contributed to improving the characteristics of the HIV epidemic, and associated fungal infection may not be as frequent as in the pre-antiretroviral time (Armstrong-James et al. 2014).

Systemic infections in people infected with this virus usually occur due to pathogens such as *P. jirovecii*, causing pneumocystosis, *C. neoformans*, the causative agent of cryptococcosis, *H. capsulatum* and *Talaromyces* (Limper et al. 2017). In developed countries with easy access to antiretroviral therapy, invasive fungal infections have decreased in recent years (Limper et al. 2017). However, in countries such as India, the clinical profile of AIDS is considered different from that seen in developed countries (Kaur et al. 2016). Living conditions in this country make the prevalence of infectious diseases higher. For example, among people with HIV-infected patients infections caused by *C. albicans, C. neoformans,* and *A. fumigatus* account for the majority of infections in this population, and the mortality rate is increasing in this country (Kaur et al. 2016). *C. albicans* is the most frequently isolated organism in these patients; however, some studies have shown that species such as *C. tropicalis, C. krusei,* and *C. glabrata* have become more prevalent than *C. albicans (*Jain et al. 2014). In African and Latin American countries, cryptococcosis, pneumocystosis, and histoplasmosis are the most frequently occurring systemic mycoses in people with HIV (Dasmasceno et al. 2013; Brown et al. 2014). These infections can be acquired by inhalation of propagules and usually present as an acute pulmonary infections (Dasmasceno et al. 2013). When these fungal infections are not controlled, systemic dissemination to the central nervous system can occur, which can lead to patient death. Other fungal species have been described that can affect HIV patients, including the pathogen *Sporothrix schenckii* (Freitas et al. 2012; Freitas et al. 2014). Several reports have shown that the infection known as sporotrichosis commonly affects the skin of these patients; however, multiorgan dissemination may occur due to virus-related immunosuppression (Lopez-Romero et al. 2011; Lopes-Bezerra et al. 2018). Data on HIV infection and sporotrichosis are still scarce; however, in the last decades there has been an increase in this infection, mainly in Brazil. Between 1999 and 2009, 21 cases of sporotrichosis in HIV patients were reported *(*Freitas et al. 2012). HIV coinfection with sporotrichosis has also been reported in countries such as the United States, Peru, Spain, Mexico, Italy, Congo, and the United Kingdom (Bustamante et al. 2009; Pinto-Almazán et al.

2023). Other fungi such as *Penicillium marneffei* have also been reported, penicilliosis is an opportunistic infection associated with HIV and occurs most frequently in Southeast Asia (Le et al. 2010).

In addition to disseminated infections, patients with this virus are prone to superficial mycotic infections, such as seborrheic dermatitis, tinea pedis, tinea corporis, and onychomycosis. However, the diagnosis of these superficial infections can be difficult due to atypical clinical manifestations (Aly and Berger 1996). It has been reported that mucocutaneous candidiasis is one of the first signs of HIV infection, and approximately 90% of patients with HIV tend to develop oropharyngeal candidiasis at some point in their disease (Samaranayake et al. 2002). *Candida* esophagitis is also of concern because it occurs in more than 10% of patients with HIV (Vazquez 2010). The main clinical manifestations of patients with this virus and who present coinfection with invasive fungal infection are weight loss, oral ulcers, fever, headache, neutropenia, and recurrent diarrhea (Kaur et al. 2016).

The identification of the fungal pathogens mentioned above is carried out based on symptomatology and laboratory tests. The most common techniques include microscopy and fungal culture on selective media, such as Sabouraud dextrose agar with chloramphenicol, blood agar, and brain heart infusion agar (Kaur et al. 2016). Blood, oral swabs, expectorate sputum, and skin/nail scrapings, among others, can also be collected. In the case of sporotrichosis, the diagnosis can be made through biological samples obtained from superficial secretions, purulent contents, aspirate of non-ulcerated gummy lesions, and biopsy samples from the edges of the lesions (Freitas et al. 2012). When disseminated infection is suspected, the etiologic agent is attempted to be recovered from sputum, urine, or cerebrospinal fluid samples (Freitas et al. 2012). It is also important to inquire about the patient's immune status, and for this purpose, a $CD4^+$ cell count should be performed, accompanied by viral load determination, to help assess the AIDS status. These results will influence the patient's treatment.

For mucocutaneous candidiasis, the first-line treatment is azoles, such as clotrimazole, fluconazole, and itraconazole. These azoles are generally effective agents in HIV-infected patients with oropharyngeal candidiasis (Vazquez 2010). However, it is frequent that these patients show clinical relapses due to immunosuppression. Previous works have reported that there are patients with mucocutaneous candidiasis who present resistance to fluconazole. For this type of case, the treatment should be changed to itraconazole or amphotericin B. However, the latter can cause greater resistance to fluconazole and side effects (Vazquez 2010). For sporotrichosis, the treatment is also based on azoles, such as itraconazole in doses of 100–200 mg and amphotericin B. These antifungals have a cure rate of 81% (Freitas et al. 2012). In the case of itraconazole, the dose is increased if the lesion worsens, and the treatment duration will depend on the clinical cure and the patient's immune status. In the case of patients with disseminated infection, the treatment to be used is amphotericin B in total doses of 1 to 2.5 mg. It is important to keep in mind that in patients with low $CD4^+$ cell counts, the amounts of antifungals administered should be modified until the cell count is >200 cells/μL (Freitas et al. 2012).

In the case of pulmonary aspergillosis, amphotericin B is still the most widely used treatment. For this fungal infection, early initiation of treatment and appropriate dosing can improve the patient's life quality (Mylonakis et al. 1998). The combination of antiretrovirals, such as tenofovir, lamivudine, and efavirenz, can also be used with the antifungal drug itraconazole at a concentration of 200 mg twice daily to treat the fungal and viral infections (Singh et al. 2021).

## *Temporary Immunodeficiency*

### Coronavirus Disease 2019 (COVID-19)

SARS-CoV-2 is considered the causative agent of the disease known as coronavirus disease 2019 (COVID-19), which rapidly spread worldwide and caused the recent pandemic (García-Carnero and Mora-Montes 2022). COVID-19 is a respiratory disease that shows as an asymptomatic, mildly symptomatic infection to severe pneumonia, leading to progressive respiratory failure, which can be treated by noninvasive tools or invasive mechanical ventilation (García-Carnero and Mora-Montes 2022). According to the World Health Organization, to date, there have been 760,360,956 confirmed cases of people infected with this virus (https://www.who.int/). Recent work has shown that disease caused by COVID-19 is associated with increased proinflammatory markers, including IL-1, IL-6, and TNFα, and decreased expression of $CD4^+$ interferon-gamma and $CD4^+$ and $CD8^+$ cells, leading to increased susceptibility to both bacterial and fungal infections (Pemán et al. 2020; Bhatt et al. 2021; Russell et al. 2021). Opportunistic invasive fungal infections have been reported in patients with COVID-19, and the main risks include leukopenia, neutropenia, immune dysregulation, poorly controlled diabetes, pulmonary disease, or other comorbidities (García-Carnero and Mora-Montes 2022). The three groups of fungal pathogens that most frequently cause coinfections in patients with COVID-19 are species of the genera *Aspergillus*, Mucorales, and *Candida* (Hoenigl et al. 2022).

Mucormycosis (caused by several species of the order Mucorales, especially *Rhizopus* spp.) associated with COVID-19 gained special attention in early 2021 in India, during the second wave of the pandemic (Chakrabarti 2021). This increase could be related to an effective association between this infection and the delta variant of SARS-CoV-2 because this variant is more contagious, which increases the risk of hospitalization, predisposition to the rhinocerebral form, as well as its capacity to affect other organs such as the pancreas (Chakrabarti 2021). Before the pandemic, mucormycosis had a mortality rate of 50%; however, mortality after the pandemic increased to 85% in India. This infection has become a major health problem, with more than 47,500 cases reported by the Indian government (Hoenigl et al. 2022). Cases of COVID-19-associated mucormycosis (CAM) have also been reported in other countries, including the United States, Egypt, Iran, Brazil, Chile, United Kingdom, France, Italy, Austria, and Mexico (García-Carnero and Mora-Montes 2022). The most frequent clinical presentations of CAM in patients with

COVID-19 are rhino-orbital-cerebral and pulmonary diseases; however, the manifestation and incidence of these vary between geographic regions (García-Carnero and Mora-Montes 2022). The main concern with this infection is the necrosis produced in the skin, which requires surgical debridement, a procedure that proves to be very invasive (Rudramurthy et al. 2021). The causative agents of mucormycosis take advantage of alterations that occur in the host due to SARS-CoV-2 infection. These include hyperglycemia, which is caused by the use of corticosteroids to treat COVID-19, as well as ACE2 receptor dysregulation, immunosuppression, hypoxia, and free iron availability due to cytokine storm. In addition, low pH, which is caused by diabetic ketoacidosis, and pulmonary changes may also contribute to fungal infection. Overexpression of GRP-78 endothelial cells and prolonged hospital stays may also increase the risk of infection (García-Carnero and Mora-Montes 2022).

Currently, CAM represents 0.3% of fungal coinfections reported by COVID-19 (Musuuza et al. 2021). Diagnosis of this fungal disease is carried out through cultures on Sabouraud medium, and microbiological and histopathological examinations of tissues from different lesions. These microbiological analyses reveal the presence of broad, nonseptate, ribbon-like hyaline hyphae; however, depending on the clinical form, the analysis may be more complicated, as processing may damage the hyphae, preventing their growth on the selective medium (Rudramurthy et al. 2021). This conventional diagnosis of CAM is time-consuming, delaying diagnosis and treatment (Hoenigl et al. 2022). Therapy for patients with CAM is similar to that of mucormycosis, and the success is directly related to early diagnosis (Rudramurthy et al. 2021). The most commonly used antifungal therapy for this infection is amphotericin B, preferably its liposomal form, which should be administered for at least 6 weeks (Rudramurthy et al. 2021). If this antifungal cannot be used, posaconazole and isavuconazole can be used as rescue therapy (Aranjani et al. 2021). In case none of these antifungals are available, itraconazole could be used as intravenous treatment, although it is not highly recommended (Rudramurthy et al. 2021). It is important to keep in mind that if there is an active SARS-CoV-2 infection, treatment may vary due to medicament interactions that may exist with antifungal, antiviral, and immune therapies (Aranjani et al. 2021). In some patients, treatment with amphotericin B has been found to generate toxicity, and some clinical isolates have been reported to show resistance (Ghosh et al. 2021).

*C. albicans* is the species most frequently found in COVID-19 infections; however, recently, the pathogen *Candida auris* has been recognized to generate resistance to antifungal drugs, especially in patients coinfected with this virus (Tsai et al. 2022). Previous studies reported that patients with COVID-19 who were in the ICU had candidiasis or candidemia, 67% of these had an infection caused by *C. auris,* and the rest of the patients' infection was caused by *C. albicans* and *C. tropicalis* (Chowdhary et al. 2020). Oxygen masks, feeding tubes, catheters, ventilation tubes, and vascular catheters can promote the transmission of candidiasis (Arastehfar et al. 2020). Some of the symptoms of COVID-19-associated candidiasis (CAC) include fever, low blood pressure, abscesses and abdominal pain, urinary tract infections, and pustular lesions (Riad et al. 2021). White lesions inside the mouth (thrush) often appear in patients with CAC, especially in those who wear dentures or prostheses

(Ghosh et al. 2021). As with CAM, patients with CAC present an excessive inflammatory reaction, which causes an increase in the level of cytokines IL-6, TNFα, INFα, and INFγ, generating a "cytokine storm" (Ragab et al. 2020). This immune response destroys tissues of blood vessels, alveoli, and capillaries, which can lead to multiorgan failure and death (Ragab et al. 2020).

The use of corticosteroids, specifically glucocorticoids, during the treatment of COVID-19 treatment in severe patients, increases the risk of developing candidemia and invasive candidiasis (Riche et al. 2020). This is due to the inhibition of TNFα secretion and changes in monocyte function, which brings about impaired pathogen recognition (Ghosh et al. 2021). Diagnosis of candidiasis in patients with CAC is performed by physical examination and by obtaining cultures from blood samples. Recently, the use of T2*Candida* nanodiagnostic has been employed, which is a molecular test that is based on magnetic resonance imaging and can distinguish between five *Candida* species (*C. albicans, C. glabrata, C. krusei, C. tropicalis*, and *C. parapsilosis* (Clancy and Nguyen 2018). Diagnosis of candidiasis in patients with COVID-19 should be prompt, as the patient's health depends on this (Ghosh et al. 2021).

The treatment for CAC is echinocandins, followed by liposomal amphotericin B, voriconazole, posaconazole, isavuconazole, and fluconazole (Arastehfar et al. 2020). Although a large number of antifungals are available for the treatment of this mycosis, many strains have been shown to generate resistance to these, resulting in CAC being difficult to control in hospital settings (Ghosh et al. 2021).

Aspergillosis is another mycosis that has been reported in people with COVID-19. This usually shows different aftermaths (Yusuf et al. 2021). The appearance of this mycosis may be related to people who have a deficient immune system and who show lung damage as a result of COVID-19 (Lai and Yu 2021). The symptoms of this infection include fever, chest pain, cough, coughing up blood, and respiratory distress (Koehler et al. 2021). The diagnosis of this infection depends on the culture of the pathogen obtained from a biopsy. In addition, biomarkers in serum or bronchoalveolar lavage fluid can be used (Garg et al. 2021). Antifungal treatment that may be effective against aspergillosis is voriconazole and isavuconazole (Arastehfar et al. 2020). However, voriconazole may have a medicament interaction with other medicines, such as remdesivir, which is given to patients with COVID-19. On the other hand, isavuconazole has a better pharmacokinetic profile and generates less toxicity (Arastehfar et al. 2020). Amphotericin B is another treatment alternative; however, it cannot be administered to patients with renal insufficiency (Ghosh et al. 2021).

## *Trauma, Burns, and Major Surgery*

### Organ Transplantation

Fungal infection is a major obstacle in solid organ transplantation (SOT). The consequences of these infections in transplant patients include long hospital stays and damage to the allograft, which can be lethal (Shoham and Marr 2012). Different studies indicate that the mortality of patients with SOT is related to these fungal infections, and the causative etiological agents are mostly species of the genus *Candida*, *Aspergillus*, and *Cryptococcus* (Pappas et al. 2010). The greatest risk is found in the small intestine, lung, liver, heart, pancreas, and kidney transplants (Pappas et al. 2010).

Different factors are related to the development of fungal infections in patients with SOT, including the patient's environmental exposure, the use of antifungal prophylaxis, and the immunosuppression level (Shoham and Marr 2012). This immunosuppression is related to antirejection therapies, ruptures of mucocutaneous barriers, leukopenia, and comorbid conditions such as malnutrition, cirrhosis, and diabetes mellitus (Fishman 2007). This deficiency in the immunological status varies depending on the type of transplant received and is important in determining specific fungal infections. In the case of invasive candidiasis, this usually appears weeks or months after the transplant and is more frequent in patients with lung and liver transplants (Neofytos et al. 2010). Invasive aspergillosis tends to appear after 6 months post-transplantation and usually affects kidney, heart, and lung transplant recipients; while cryptococcosis tends to occur between 2 to 5 years after transplantation (Neofytos et al. 2010). Transplanted organs can act as reservoirs for pathogenic fungi, including species of *Aspergillus*, *Candida, Histoplasma capsulatum, C. neoformans, Coccidioides immitis*, and *Scedosporium apiospermun* (Shoham and Marr 2012). Before carrying out the transplant, it is important to take into account that the donor must not present signs of fungal infections since this is considered a contraindication (Fischer and Avery 2009).

Patients receiving SOT have a higher risk of acquiring invasive candidiasis, which is considered the most common mycosis in this type of patient (van Hal et al. 2009). *C. albicans* is known to be the most common etiologic agent causing this infection, with an incidence of approximately 50%, *C. parapsilosis* is known to be associated with permanent medical device infections and is considered an important pathogen in people with SOT (Shoham and Marr 2012). The species *C. glabrata* and *C. krusei* are also important pathogens; however, both species are more associated with those who have received antifungal treatment before transplantation (Neofytos et al. 2010). The development of invasive candidiasis depends on the deterioration of the patient's immune response, and the virulence of the organism. Abdominal organ transplant patients are often the most affected by these pathogens and have multiple risk factors for contracting invasive candidiasis. These include the presence of central venous catheters, abdominal surgery, the use of corticosteroids, and treatment with broad-spectrum antibiotics. It is possible that during the surgical intervention to carry out the SOT, the different *Candida* spp. can be transmitted, and these species can end up contaminating the organ. In cases of kidney transplantation, the infection could reach the urinary tract or affect the wound site, causing organ loss (Albano et al. 2009). Mortality from invasive candidiasis in people with SOT reaches a percentage between 20 and 40% (Husain et al. 2003). The diagnosis for the different *Candida* spp. includes the growth of colonies in selective media such as CHROMagar, metabolic tests, and fluorescent in situ hybridization (Shoham and Marr 2012). The treatment of invasive candidiasis in patients with SOT is similar to that of other patients, it is necessary to use some echinocandins as initial treatment, which includes caspofungin, micafungin, or anidulafungin (Pappas et al. 2009). Infection with *C. parapsilosis* and *C. guilliermondii* should be treated with alternative agents, as both species are resistant to echinocandins (Pappas et al. 2009). Other antifungals such as amphotericin B or voriconazole can be used; however, these are often highly toxic. Once the fungus has been eliminated from the bloodstream and the symptoms of candidiasis have disappeared, treatment should be continued for a further two week period (Pappas et al. 2009).

In the case of infection caused by *Aspergillus* in patients with SOT, this usually occurs almost exclusively by inhalation of conidia from the environment. The infection involves the respiratory tract and paranasal sinuses. The most common species is *A. fumigatus*; however, some cases may be related to *Aspergillus flavus, Aspergillus niger,* and *Aspergillus terreus* (Walsh et al. 2008). The infection usually arises from the progression of a previously inactive subclinical process, or it may be due to an infection acquired after transplantation (Raviv et al. 2007). During invasive aspergillosis, the serious development of the infection is marked by the immunological status of the patient, the type of organ transplanted, and the intensity of exposure (Gavalda et al. 2005). In lung transplant patients, invasive aspergillosis accounts for 50% of fungal diseases (Pappas et al. 2010). The diagnosis for this infection is usually complicated, and histological analysis of infection and culture are often required. Smear or tissue specimens can be stained with periodic acid-Schiff or Calcofluor white (Hope et al. 2005). Lung radiography is often used for early diagnosis of infection (Hope et al. 2005). Other techniques commonly used to diagnose infection are PCR assays and galactomannan tests, with serum galactomannan being a widely used assay (Clancy et al. 2007). The first-line treatment for this infection is voriconazole, and amphotericin B can be used as an alternative agent in those patients who do not tolerate voriconazole (Baddley et al. 2010). In some cases, although the infection is already being treated with antifungals, complementary surgical debridement is recommended.

Cryptococcosis is the third most common cause of mycosis in patients with SOT, it represents 7–8% of infections (Neofytos et al. 2010; Pappas et al. 2010). The most common species is *C. neoformans*; however, *Cryptococcus gattii* can also cause infection (Harris et al. 2011). Risk factors for this infection include the use of corticosteroids and agents that deplete T cells (Silveira et al. 2007). The most common sites of infection are the lungs and the central nervous system; however, in some cases, the infection can spread and affect multiple organs (Singh et al. 2007). The diagnosis of this infection is based on radiographic findings, followed by the detection of the serum cryptococcal antigen; however, this test only works in those patients who have a high fungal load. Gram stain with India ink and culture of cerebrospinal fluid usually work for the diagnosis of these fungal agents (Singh et al. 2007). The treatment varies according to the site and severity of the infection, for those patients with severe neurological and pulmonary infection, the treatment is amphotericin B, and in some cases, flucytosine can be added. Therapy can be extended to 4 weeks, after this time, once the patient has been stabilized, treatment with fluconazole can be started at a dose of 400 to 800 mg per day for 8 weeks. The dose is then decreased to 200 mg for 6 to 8 months (Singh et al. 2007).

## *Immunosuppressive Medications*

Patients with autoimmune diseases such as rheumatoid arthritis, psoriasis, or autoimmune hemolytic anemia, as well as transplant patients, are often subjected to long treatments with drugs that cause negative effects on the immune system (Schmidt et al. 2019). Glucocorticosteroids, such as prednisolone or methylprednisolone are immunosuppressive drugs that suppress the phagocytic function of monocytes and neutrophils, alter antigen presentation, T-cell function, and secretion of proinflammatory cytokines (Tramsen et al. 2014). The medicine mycophenolate mofetil, which is used in transplant patients, affects the recruitment of monocytes and lymphocytes at sites of inflammation and can induce apoptosis of activated T lymphocytes (Allison 2005). Another commonly used immunosuppressive agent is the calcineurin inhibitor cyclosporin A, which is responsible for inhibiting the activation of T cells (Flores et al. 2019). The fungal species that appear most frequently in patients with these immunosuppressive treatments are *A. fumigatus, C. albicans,* and *C. neoformans*.

In patients with hematologic malignancies receiving chemotherapy or hematopoietic stem cell transplantation, infections caused by *Aspergillus* spp. are the most common (Khan et al. 2010). The chemotherapy used for the treatment of cancer usually has effects on the humoral and cellular immunity of the patients, causing loss of function of T lymphocytes, monocytes, macrophages, neutrophils, and mucous membranes (Vento and Cainelli 2003). A study conducted in China with 4192 patients with different types of cancer who received chemotherapy indicated that 407 patients had invasive fungal infections associated with chemotherapy (Sun et al. 2015). The incidence of invasive fungal infections was higher in patients with certain hematological diagnoses, such as myelodysplastic syndrome and acute hyperleukocytic leukemia. Furthermore, patients older than 60 years were more predisposed to these infections than young children. The genera with a higher incidence of appearance were *Candida* and *Aspergillus* (Sun et al. 2015). For the treatment of these mycoses in patients receiving chemotherapy, the first option is the use of triazoles, specifically fluconazole, itraconazole, and voriconazole. Some cases must be managed with intravenous therapy. In those cases where treatment with triazoles was not satisfactory, echinocandins and polyenes such as amphotericin B

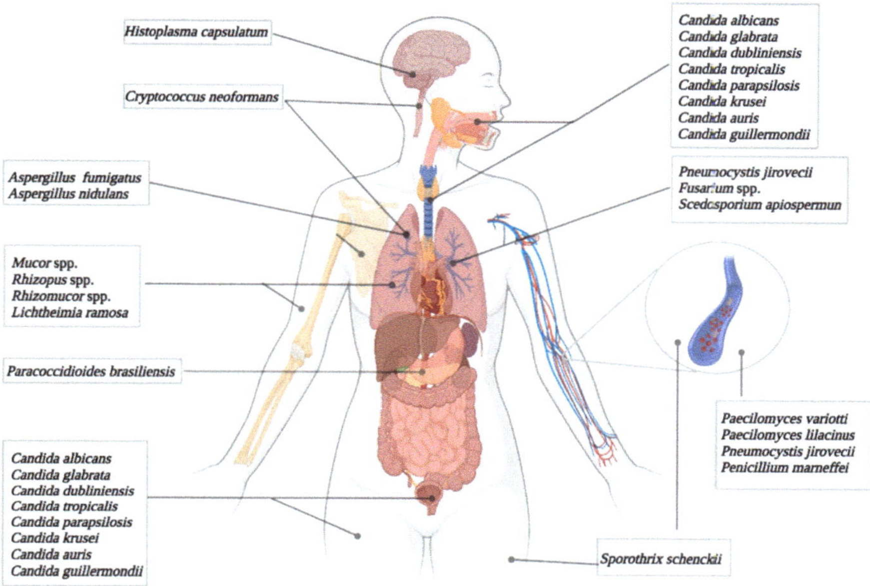


**Fig. 2** Graphic representation of the human body organs and the corresponding species of pathogenic fungi that can cause infections in individuals with primary and secondary immunodeficiencies

and caspofungin were used. The doses varied depending on the severity of the infection and the immunological status of the patient (Sun et al. 2015). Other fungal genera, such as *Fusarium* and *Trichosporon*, have been reported to appear in patients receiving chemotherapy treatment; however, low frequency (Warnock 1998).

Finally, Fig. 2 represents the location of the different species of pathogenic fungi that affect patients with primary and secondary immunodeficiencies.

## Conclusions

The presence of primary and secondary immunodeficiencies significantly increases the risk of serious and life-threatening fungal infections. These infections can be difficult to diagnose and treat, which can lead to increased morbidity and mortality in affected patients. Physicians must consider fungal infections as a possible complication in patients with immunodeficiencies, and that early and accurate evaluation be performed for diagnosis and treatment.

Although important advances have been made in the diagnosis and prevention of these infections, there is still much to be done. New prophylactic treatments, such as antifungal vaccines, monoclonal antibodies, and quorum-sensing inhibitors, show great potential in preventing these infections in immunocompromised patients. However, more research is still required to assess its efficacy and safety.

Finally, proper management of these fungal infections in patients with immunodeficiencies remains a challenge for clinicians, and an interdisciplinary approach and individualized care for each patient are needed to achieve a better life quality and survival.

## References

Abbott KC, Hypolite I, Poropatich RK, Hshieh P, Cruess D, Hawkes CA, Agodoa LY, Keller RA (2001) Hospitalizations for fungal infections after renal transplantation in the United States. Transpl Infect Dis 3(4):203–211

Ahlin A, Lärfars G, Elinder G, Palmblad J, Gyllenhammar H (1999) Gamma interferon treatment of patients with chronic granulomatous disease is associated with augmented production of nitric oxide by polymorphonuclear neutrophils. Clin Diagn Lab Immunol 6(3):420–424

Ahonen P (1985) Autoimmune polyendocrinopathy—candidosis–ectodermal dystrophy (APECED): autosomal recessive inheritance. Clin Genet 27(6):535–542

Akpan A, Morgan R (2002) Oral candidiasis. Postgrad Med J 78(922):455–459

Albano L, Bretagne S, Mamzer-Bruneel MF, Kacso I, Desnos-Ollivier M, Guerrini P, Le Luong T, Cassuto E, Dromer F, Lortholary O (2009) Evidence that graft-site candidiasis after kidney transplantation is acquired during organ recovery: a multicenter study in France. Clin Infect Dis 48(2):194–202

Alcántara-Montiel JC, Vega-Torres BI (2016) Hyper-IgE syndrome. Lessons from function and defects of STAT-3 or DOCK-8. Rev Alerg Mex 63(4):385–396

Allison AC (2005) Mechanisms of action of mycophenolate mofetil. Lupus 14(Suppl 1):s2–s8

Almeida F, Rodrigues ML, Coelho C (2019) The still underestimated problem of fungal diseases worldwide. Front Microbiol 10:214

Aly R, Berger T (1996) Common superficial fungal infections in patients with AIDS. Clin Infect Dis 22(Suppl 2):S128–S132

Antachopoulos C (2010) Invasive fungal infections in congenital immunodeficiencies. Clin Microbiol Infect 16(9):1335–1342

Antachopoulos C, Walsh TJ, Roilides E (2007) Fungal infections in primary immunodeficiencies. Eur J Pediatr 166(11):1099–1117

Aranjani JM, Manuel A, Abdul Razack HI, Mathew ST (2021) COVID-19-associated mucormycosis: evidence-based critical review of an emerging infection burden during the pandemic's second wave in India. PLoS Negl Trop Dis 15(11):e0009921

Arastehfar A, Carvalho A, van de Veerdonk FL, Jenks JD, Koehler P, Krause R, Cornely OA, Perlin DS, Lass-Flörl C, Hoenigl M (2020) COVID-19 associated pulmonary aspergillosis (CAPA)-from immunology to treatment. J Fungi (Basel) 6(2):91

Armstrong-James D, Meintjes G, Brown GD (2014) A neglected epidemic: fungal infections in HIV/AIDS. Trends Microbiol 22(3):120–127

Axelrod H, Adams M (2019) Biologic agents and secondary immune deficiency. Pediatr Clin N Am 66(5):1007–1020

Baddley JW, Andes DR, Marr KA, Kontoyiannis DP, Alexander BD, Kauffman CA, Oster RA, Anaissie EJ, Walsh TJ, Schuster MG, Wingard JR, Patterson TF, Ito JI, Williams OD, Chiller T, Pappas PG, t. T. A. I. S. Network (2010) Factors associated with mortality in transplant patients with invasive aspergillosis. Clin Infect Dis 50(12):1559–1567

Badiee P, Hashemizadeh Z (2014) Opportunistic invasive fungal infections: diagnosis & clinical management. Indian J Med Res 139(2):195–204

Bassetti M, Garnacho-Montero J, Calandra T, Kullberg B, Dimopoulos G, Azoulay E, Chakrabarti A, Kett D, Leon C, Ostrosky-Zeichner L, Sanguinetti M, Timsit JF, Richardson MD, Shorr A, Cornely OA (2017) Intensive care medicine research agenda on invasive fungal infection in critically ill patients. Intensive Care Med 43(9):1225–1238

Beisel WR (1996) Nutrition and immune function: overview. J Nutrit 126:2611S–2615S

Bhatt K, Agolli A, Patel MH, Garimella R, Devi M, Garcia E, Amin H, Domingue C, Guerra Del Castillo R, Sanchez-Gonzalez M (2021) High mortality co-infections of COVID-19 patients: mucormycosis and other fungal infections. Discoveries (Craiova) 9(1):e126

Björses P, Aaltonen J, Horelli-Kuitunen N, Yaspo ML, Peltonen L (1998) Gene defect behind APECED: a new clue to autoimmunity. Hum Mol Genet 7(10):1547–1553

Bongomin F, Gago S, Oladele RO, Denning DW (2017) Global and multi-national prevalence of fungal diseases-estimate precision. J Fungi (Basel) 3(4):57

Bonilla FA, Khan DA, Ballas ZK, Chinen J, Frank MM et al (2015) Practice parameter for the diagnosis and management of primary immunodeficiency. J Allergy Clin Immunol 136(5):1186–1205.e1181–1178

Bourke CD, Berkley JA, Prendergast AJ (2016) Immune dysfunction as a cause and consequence of malnutrition. Trends Immunol 37(6):386–398

Bousfiha AA, Jeddane L, Ailal F, Benhsaien I, Mahlaoui N, Casanova J-L, Abel L (2013) Primary immunodeficiency diseases worldwide: more common than generally thought. J Clin Immunol 33(1):1–7

Bousfiha A, Jeddane L, Al-Herz W, Ailal F, Casanova JL et al (2015) The 2015 IUIS phenotypic classification for primary immunodeficiencies. J Clin Immunol 35(8):727–738

Boyle JM, Buckley RH (2007) Population prevalence of diagnosed primary immunodeficiency diseases in the United States. J Clin Immunol 27(5):497–502

Brath H, Paldánius PM, Bader G, Kolaczynski WM, Nilsson PM (2016) Differences in glycemic control across world regions: a post-hoc analysis in patients with type 2 diabetes mellitus on dual antidiabetes drug therapy. Nutr Diabetes 6(7):e217

Brown GD, Denning DW, Gow NA, Levitz SM, Netea MG, White TC (2012) Hidden killers: human fungal infections. Sci Transl Med 4(165):165rv113

Brown GD, Meintjes G, Kolls JK, Gray C, Horsnell W et al (2014) AIDS-related mycoses: the way forward. Trends Microbiol 22(3):107–109

Bustamante B, Lama JR, Mosquera C, Soto L (2009) Sporotrichosis in human immunodeficiency virus infected peruvian patients: two case reports and literature review. Infect Dis Clin Practice 17(2):78–83

Cai CW, Sereti I (2021) Residual immune dysfunction under antiretroviral therapy. Semin Immunol 51:101471

Calvet HM, Yoshikawa TT (2001) Infections in diabetes. Infect Dis Clin N Am 15(2):407–421. viii

Capalbo D, De Martino L, Giardino G, Di Mase R, Di Donato I, Parenti G, Vajro P, Pignata C, Salerno M (2012) Autoimmune polyendocrinopathy candidiasis ectodermal dystrophy: insights into genotype-phenotype correlation. Int J Endocrinol 2012:353250

Cardinez C, Miraghazadeh B, Tanita K, da Silva E, Hoshino A et al (2018) Gain-of-function IKBKB mutation causes human combined immune deficiency. J Exp Med 215(11):2715–2724

Chakrabarti S (2021) Mental health in hospitalised COVID 19 patients in quarantine during second wave in a South Indian private teaching hospital. J Multidiscip Healthc 14:2777–2789

Chandesris MO, Melki I, Natividad A, Puel A, Fieschi C et al (2012) Autosomal dominant STAT3 deficiency and hyper-IgE syndrome: molecular, cellular, and clinical features from a French national survey. Medicine (Baltimore) 91(4):e1–e19

Chinen J, Shearer WT (2010) Secondary immunodeficiencies, including HIV infection. J Allergy Clin Immunol 125(2 Suppl 2):S195–S203

Chowdhary A, Tarai B, Singh A, Sharma A (2020) Multidrug-resistant *Candida auris* infections in critically ill coronavirus disease patients, India, april-july 2020. Emerg Infect Dis 26(11):2694–2696

Clancy CJ, Nguyen MH (2018) Diagnosing invasive candidiasis. J Clin Microbiol 56(5):e01909–e01917

Clancy CJ, Jaber RA, Leather HL, Wingard JR, Staley B et al (2007) Bronchoalveolar lavage galactomannan in diagnosis of invasive pulmonary aspergillosis among solid-organ transplant recipients. J Clin Microbiol 45(6):1759–1765

Colman S, Giusiano G, Colman C, Sosa M, Rojas F (2022) Hepatic failure and malnutrition as predisposing factors of cutaneous mucormycosis in a pediatric patient. Med Mycol Case Rep 35:26–29

Conti HR, Shen F, Nayyar N, Stocum E, Sun JN et al (2009) Th17 cells and IL-17 receptor signaling are essential for mucosal host defense against oral candidiasis. J Exp Med 206(2):299–311

Corvilain E, Casanova JL, Puel A (2018) Inherited CARD9 deficiency: invasive disease caused by ascomycete fungi in previously healthy children and adults. J Clin Immunol 38(6):656–693

Corzo-León DE, Chora-Hernández LD, Rodríguez-Zulueta AP, Walsh TJ (2018) Diabetes mellitus as the major risk factor for mucormycosis in Mexico: epidemiology, diagnosis, and outcomes of reported cases. Med Mycol 56(1):29–43

Cuvelier GDE, Rubin TS, Junker A, Sinha R, Rosenberg AM, Wall DA, Schroeder ML (2019) Clinical presentation, immunologic features, and hematopoietic stem cell transplant outcomes for IKBKB immune deficiency. Clin Immunol 205:138–147

Dasmasceno LS, Novaes AR Jr, Alencar CH, Lima DT, Sidrim JJ, Gonçalves MV, de Mesquita JR, Leitão Tdo M (2013) Disseminated histoplasmosis and aids: relapse and late mortality in endemic area in North-Eastern Brazil. Mycoses 56(5):520–526

Denning DW, Pleuvry A, Cole DC (2013) Global burden of chronic pulmonary aspergillosis complicating sarcoidosis. Eur Respir J 41(3):621–626

Desai A, Schmitt-Hoffmann AH, Mujais S, Townsend R (2016) Population pharmacokinetics of isavuconazole in subjects with mild or moderate hepatic impairment. Antimicrob Agents Chemother 60(5):3025–3031

Drewniak A, Gazendam RP, Tool AT, van Houdt M, Jansen MH et al (2013) Invasive fungal infection and impaired neutrophil killing in human CARD9 deficiency. Blood 121(13):2385–2392

Drummond RA, Franco LM, Lionakis MS (2018) Human CARD9: a critical molecule of fungal immune surveillance. Front Immunol 9:1836

Emmanuel et al (1999) https://doi.org/10.1128/IAI.67.6.3047-3050.1999

Espinoza G, Butte K, Palma V, Norambuena X, Quezada A (2015) Enfermedad granulomatosa crónica: tres casos clínicos con diferentes formas de presentación. Rev Chil Ped 86(2):112–116

Field et al (2002) https://pubmed.ncbi.nlm.nih.gov/11781377/

Fischer SA, Avery RK (2009) Screening of donor and recipient prior to solid organ transplantation. Am J Transplant 9 Suppl 4(Suppl 4):S7–S18

Fishman JA (2007) Infection in solid-organ transplant recipients. New Engl J Med 357(25):2601–2614

Flores C, Fouquet G, Moura IC, Maciel TT, Hermine O (2019) Lessons to learn from low-dose cyclosporin-A: a new approach for unexpected clinical applications. Front Immunol 10:588

Freitas DF, de Siqueira Hoagland B, do Valle AC, Fraga BB, de Barros MB et al (2012) Sporotrichosis in HIV-infected patients: report of 21 cases of endemic sporotrichosis in Rio de Janeiro, Brazil. Med Mycol 50(2):170–178

Freitas DF, Valle AC, da Silva MB, Campos DP, Lyra MR, de Souza RV, Veloso VG, Zancopé-Oliveira RM, Bastos FI, Galhardo MC (2014) Sporotrichosis: an emerging neglected opportunistic infection in HIV-infected patients in Rio de Janeiro, Brazil. PLoS Negl Trop Dis 8(8):e3110

García-Carnero LC, Mora-Montes HM (2022) Mucormycosis and COVID-19-associated mucormycosis: insights of a deadly but neglected mycosis. J Fungi (Basel) 8(5):445

Gandhi BV, Bahadur MM, Dodeja H, Aggrwal V, Thamba A, Mali M (2005) Infecciones fúngicas sistémicas en enfermedades renales. JPostgrad Med [serie en línea] 2005 [citado el 29 de abril de 2023];51:30–36

Garg D, Muthu V, Sehgal IS, Ramachandran R, Kaur H, Bhalla A, Puri GD, Chakrabarti A, Agarwal R (2021) Coronavirus disease (Covid-19) associated mucormycosis (CAM): case report and systematic review of literature. Mycopathologia 186(2):289–298

Gavalda J, Len O, San Juan R, Aguado JM, Fortun J et al (2005) Risk factors for invasive aspergillosis in solid-organ transplant recipients: a case-control study. Clin Infect Dis 41(1):52–59

Ghosh A, Sarkar A, Paul P, Patel P (2021) The rise in cases of mucormycosis, candidiasis and aspergillosis amidst COVID19. Fungal Biol Rev 38:67–91

Glocker EO, Hennigs A, Nabavi M, Schaeffer AA, Woellner C et al (2009) A homozygous CARD9 mutation in a family with susceptibility to fungal infections. N Engl J Med 361(18):1727–1735

Gómez-Gaviria M, Ramírez-Sotelo U, Mora-Montes HM (2023) Non-*albicans Candida* species: immune response, evasion mechanisms, and new plant-derived alternative therapies. J Fungi 9(1):11

Guinea J (2014) Global trends in the distribution of *Candida* species causing candidemia. Clin Microbiol Infect 20(Suppl 6):5–10

Gürsoy S, Koçkar T, Atik SU, Önal Z, Önal H, Adal E (2018) Autoimmunity and intestinal colonization by *Candida albicans* in patients with type 1 diabetes at the time of the diagnosis. Korean J Pediatr 61(7):217–220

Hanachi M, Bohem V, Bemer P, Kayser N, de Truchis P, Melchior JC (2018) Negative role of malnutrition in cell-mediated immune response: pneumocystis jirovecii pneumonia (PCP) in a severely malnourished, HIV-negative patient with anorexia nervosa. Clin Nutr ESPEN 25:163–165

Harris JR, Lockhart SR, Debess E, Marsden-Haug N, Goldoft M, Wohrle R, Lee S, Smelser C, Park B, Chiller T (2011) *Cryptococcus gattii* in the United States: clinical aspects of infection with an emerging pathogen. Clin Infect Dis 53(12):1188–1195

Heino M, Peterson P, Kudoh J, Shimizu N, Antonarakis SE, Scott HS, Krohn K (2001) APECED mutations in the autoimmune regulator (AIRE) gene. Hum Mutat 18(3):205–211

Hoenigl M, Prattes J, Spiess B, Wagner J, Prueller F et al (2014) Performance of galactomannan, beta-d-glucan, *Aspergillus* lateral-flow device, conventional culture, and PCR tests with bronchoalveolar lavage fluid for diagnosis of invasive pulmonary aspergillosis. J Clin Microbiol 52(6):2039–2045

Hoenigl M, Seidel D, Sprute R, Cunha C, Oliverio M, Goldman GH, Ibrahim AS, Carvalho A (2022) COVID-19-associated fungal infections. Nat Microbiol 7(8):1127–1140

Hope WW, Walsh TJ, Denning DW (2005) Laboratory diagnosis of invasive aspergillosis. Lancet Infect Dis 5(10):609–622

Hsu AP, Sampaio EP, Khan J, Calvo KR, Lemieux JE et al (2011) Mutations in GATA2 are associated with the autosomal dominant and sporadic monocytopenia and mycobacterial infection (MonoMAC) syndrome. Blood 118(10):2653–2655

Husain S, Tollemar J, Dominguez EA, Baumgarten K, Humar A, Paterson DL, Wagener MM, Kusne S, Singh N (2003) Changes in the spectrum and risk factors for invasive candidiasis in liver transplant recipients: prospective, multicenter, case-controlled study. Transplantation 75(12):2023–2029

Husebye ES, Perheentupa J, Rautemaa R, Kämpe O (2009) Clinical manifestations and management of patients with autoimmune polyendocrine syndrome type I. J Intern Med 265(5):514–529

Ilkit M, Durdu M, Karakaş M (2012) Majocchi's granuloma: a symptom complex caused by fungal pathogens. Med Mycol 50(5):449–457

Iseki M, Anzo M, Yamashita N, Matsuo N (1994) Hyper-IgM immunodeficiency with disseminated cryptococcosis. Acta Paediatr 83(7):780–782

Jabra-Rizk MA, Falkler WA Jr, Enwonwu CO, Onwujekwe DI Jr, Merz WG, Meiller TF (2001) Prevalence of yeast among children in Nigeria and the United States. Oral Microbiol Immunol 16(6):383–385

Jain S, Kumar Singh A, Pal Singh R, Bajaj JK, Damle AS (2014) Spectrum of opportunistic fungal infections in HIV-infected patients and their correlation with CD4+ counts in western India. J Med Microbiol Infect Dis 2(1):19–22

Jawale CV, Ramani K, Li DD, Coleman BM, Oberoi RS et al (2020) Restoring glucose uptake rescues neutrophil dysfunction and protects against systemic fungal infection in mouse models of kidney disease. Sci Transl Med 12(548)
Jirapongsananuruk O, Luangwedchakarn V, Niemela JE, Pacharn P, Visitsunthorn N, Thepthai C, Vichyanond P, Piboonpocanun S, Fleisher TA (2012) Cryptococcal osteomyelitis in a child with a novel compound mutation of the IL12RB1 gene. Asian Pac J Allergy Immunol 30(1):79–82
Jongco AM, Gough JD, Sarnataro K, Rosenthal DW, Moreau J, Ponda P, Bonagura VR (2014) X-linked agammaglobulinemia presenting as polymicrobial pneumonia, including *pneumocystis jirovecii*. Ann Allergy Asthma Immunol 112(1):74–75.e72
José RJ, Brown JS (2016) Opportunistic bacterial, viral and fungal infections of the lung. Medicine (Abingdon) 44(6):378–383
Juilland M, Thome M (2018) Holding all the CARDs: how MALT1 controls CARMA/CARD-dependent signaling. Front Immunol 9:1927
Jung S, Gies V, Korganow AS, Guffroy A (2020) Primary immunodeficiencies with defects in innate immunity: focus on orofacial manifestations. Front Immunol 11:1065
Kauer J, Vogt F, Hagelstein I, Hoerner S, Maerklin M et al (2021) CD18 antibody application blocks unwanted off-target T cell activation caused by bispecific antibodies. Cancers (Basel) 13(18)
Kaur R, Dhakad MS, Goyal R, Bhalla P, Dewan R (2016) Spectrum of opportunistic fungal infections in HIV/AIDS patients in tertiary care hospital in India. Can J Infect Dis Med Microbiol 2016:2373424
Khan MSA, Ahmad I, Aqil F, Owais M, Shahid M, Musarrat J (2010) Virulence and pathogenicity of fungal pathogens with special reference to *Candida albicans*. In: Ahmad I, Owais M, Shahid M, Aqil F (eds) Combating fungal infections: problems and remedy. Springer Berlin Heidelberg, Berlin, Heidelberg, pp 21–45
Kisand K, Peterson P (2015) Autoimmune polyendocrinopathy candidiasis ectodermal dystrophy. J Clin Immunol 35(5):463–478
Klingspor L, Stitzing G, Johansen K, Murtaza A, Holmberg K (1993) Infantile diarrhoea and malnutrition associated with *Candida* in a developing community. Mycoses 36(1–2):19–24
Koehler P, Bassetti M, Chakrabarti A, Chen SCA, Colombo AL et al (2021) Defining and managing COVID-19-associated pulmonary aspergillosis: the 2020 ECMM/ISHAM consensus criteria for research and clinical guidance. Lancet Infect Dis 21(6):e149–e162
Lai CC, Yu WL (2021) COVID-19 associated with pulmonary aspergillosis: a literature review. J Microbiol Immunol Infect 54(1):46–53
Lanternier F, Cypowyj S, Picard C, Bustamante J, Lortholary O, Casanova JL, Puel A (2013) Primary immunodeficiencies underlying fungal infections. Curr Opin Pediatr 25(6):736–747
Lao M, Li C, Li J, Chen D, Ding M, Gong Y (2020) Opportunistic invasive fungal disease in patients with type 2 diabetes mellitus from southern China: clinical features and associated factors. J Diabetes Investig 11(3):731–744
Le T, Huu Chi N, Kim Cuc NT, Manh Sieu TP et al (2010) AIDS-associated *Penicillium marneffei* infection of the central nervous system. Clin Infect Dis 51(12):1458–1462
Li Y, Fang W, Jiang W, Hagen F, Liu J, Zhang L et al (2017) Cryptococcosis in patients with diabetes mellitus II in mainland China: 1993–2015. Mycoses 60(11):706–713
Limper AH, Adenis A, Le T, Harrison TS (2017) Fungal infections in HIV/AIDS. Lancet Infect Dis 17(11):e334–e343
Lionakis MS (2019) Primary immunodeficiencies and invasive fungal infection: when to suspect and how to diagnose and manage. Curr Opin Infect Dis 32(6):531–537
Lopes-Bezerra LM, Mora-Montes HM, Zhang Y, Nino-Vega G, Rodrigues AM et al (2018) Sporotrichosis between 1898 and 2017: the evolution of knowledge on a changeable disease and on emerging etiological agents. Med Mycol 56(suppl_1):S126–S143
López LF, Valencia Y, Tobón AM, Velásquez O, Santa CD, Cáceres DH, Restrepo Á, Cano LE (2016) Childhood histoplasmosis in Colombia: clinical and laboratory observations of 45 patients. Med Mycol 54(7):677–683

Lopez-Romero E, del Rocío Reyes-Montes M, Perez-Torres A, Ruiz-Baca E, Villagomez-Castro JC et al (2011) *Sporothrix schenckii* complex and sporotrichosis, an emerging health problem. Future Microbiol 6(1):85–102

Lu HY, Turvey SE (2021) Human MALT1 deficiency and predisposition to infections. Curr Opin Immunol 72:1–12

Lu HY, Bauman BM, Arjunaraja S, Dorjbal B, Milner JD, Snow AL, Turvey SE (2018) The CBM-opathies-a rapidly expanding spectrum of human inborn errors of immunity caused by mutations in the CARD11-BCL10-MALT1 complex. Front Immunol 9:2078

McKinnon ML, Rozmus J, Fung SY, Hirschfeld AF, Del Bel KL et al (2014) Combined immunodeficiency associated with homozygous MALT1 mutations. J Allergy Clin Immunol 133(5):1458–1462, 1462.e1451–1457

Mendes RP, Cavalcante RS, Marques SA, Marques MEA, Venturini J et al (2017) Paracoccidioidomycosis: current perspectives from Brazil. Open Microbiol J 11:224–282

Méndez-Samperio P, Trejo-Echeverria A, Ayala-Verdin H (1998) Regulation of interleukin-12 production in human cells stimulated with *Mycobacterium bovis* BCG. Cell Immunol 189(1):25–30

Modell V, Orange JS, Quinn J, Modell F (2018) Global report on primary immunodeficiencies: 2018 update from the Jeffrey Modell Centers Network on disease classification, regional trends, treatment modalities, and physician reported outcomes. Immunol Res 66(3):367–380

Moraes-Vasconcelos D, Grumach AS, Yamaguti A, Andrade ME, Fieschi C et al (2005) *Paracoccidioides brasiliensis* disseminated disease in a patient with inherited deficiency in the beta1 subunit of the interleukin (IL)-12/IL-23 receptor. Clin Infect Dis 41(4):e31–e37

Morton JR, Mihalas LS, Leung P, Strieder DJ (1980) Corticosteroids and malnutrition: *Aspergillus* lung abscess in an asthmatic child. Chest 78(4):667–669

Mousallem T, Yang J, Urban TJ, Wang H, Adeli M, Parrott RE, Roberts JL, Goldstein DB, Buckley RH, Zhong XP (2014) A nonsense mutation in IKBKB causes combined immunodeficiency. Blood 124(13):2046–2050

Musuuza JS, Watson L, Parmasad V, Putman-Buehler N, Christensen L, Safdar N (2021) Prevalence and outcomes of co-infection and superinfection with SARS-CoV-2 and other pathogens: a systematic review and meta-analysis. PLoS One 16(5):e0251170

Mylonakis E, Barlam TF, Flanigan T, Rich JD (1998) Pulmonary aspergillosis and invasive disease in AIDS: review of 342 cases. Chest 114(1):251–262

Nagappan R, Collins JF, Lee WT (1992) Fungal peritonitis in continuous ambulatory peritoneal dialysis–the Auckland experience. Am J Kidney Dis 20(5):492–496

Neofytos D, Fishman JA, Horn D, Anaissie E, Chang CH, Olyaei A, Pfaller M, Steinbach WJ, Webster KM, Marr KA (2010) Epidemiology and outcome of invasive fungal infections in solid organ transplant recipients. Transpl Infect Dis 12(3):220–229

Nielsen C, Jakobsen MA, Larsen MJ, Müller AC, Hansen S, Lillevang ST, Fisker N, Barington T (2014) Immunodeficiency associated with a nonsense mutation of IKBKB. J Clin Immunol 34(8):916–921

Ouederni M, Sanal O, Ikinciogullari A, Tezcan I, Dogu F et al (2014) Clinical features of candidiasis in patients with inherited interleukin 12 receptor β1 deficiency. Clin Infect Dis 58(2):204–213

Paillaud E, Merlier I, Dupeyron C, Scherman E, Poupon J, Bories PN (2004) Oral candidiasis and nutritional deficiencies in elderly hospitalised patients. Br J Nutr 92(5):861–867

Pannicke U, Baumann B, Fuchs S, Henneke P, Rensing-Ehl A et al (2013) Deficiency of innate and acquired immunity caused by an IKBKB mutation. N Engl J Med 369(26):2504–2514

Pappas PG, Kauffman CA, Andes D, Benjamin DK Jr, Calandra TF et al (2009) Clinical practice guidelines for the management candidiasis: 2009 update by the Infectious Diseases Society of America. Clin Infect Dis 48(5):503–535

Pappas PG, Alexander BD, Andes DR, Hadley S, Kauffman CA et al (2010) Invasive fungal infections among organ transplant recipients: results of the Transplant-Associated Infection Surveillance Network (TRANSNET). Clin Infect Dis 50(8):1101–1111

Pemán J, Ruiz-Gaitán A, García-Vidal C, Salavert M, Ramírez P et al (2020) Fungal co-infection in COVID-19 patients: should we be concerned? Rev Iberoam Micol 37(2):41–46

Perheentupa J (2006) Autoimmune polyendocrinopathy-candidiasis-ectodermal dystrophy. J Clin Endocrinol Metab 91(8):2843–2850
Picard C, Al-Herz W, Bousfiha A, Casanova JL, Chatila T et al (2015) Primary immunodeficiency diseases: an update on the classification from the International Union of Immunological Societies Expert Committee for Primary Immunodeficiency 2015. J Clin Immunol 35(8):696–726
Picard C, Bobby Gaspar H, Al-Herz W, Bousfiha A et al (2018) International Union of Immunological Societies: 2017 primary immunodeficiency diseases committee report on inborn errors of immunity. J Clin Immunol 38(1):96–128
Pinto-Almazán R, Sandoval-Navarro KA, Damián-Magaña EJ, Arenas R, Fuentes-Venado CE et al (2023) Relationship of sporotrichosis and infected patients with HIV-AIDS: an actual systematic review. J Fungi (Basel) 9(4):396
Prasad KN, Prasad N, Gupta A, Sharma RK, Verma AK, Ayyagari A (2004) Fungal peritonitis in patients on continuous ambulatory peritoneal dialysis: a single centre Indian experience. J Infect 48(1):96–101
Puel A, Doeffinger R, Natividad A, Chrabieh M, Barcenas-Morales G et al (2010) Autoantibodies against IL-17A, IL-17F, and IL-22 in patients with chronic mucocutaneous candidiasis and autoimmune polyendocrine syndrome type I. J Exp Med 207(2):291–297
Puel A, Cypowyj S, Maródi L, Abel L, Picard C, Casanova JL (2012) Inborn errors of human IL-17 immunity underlie chronic mucocutaneous candidiasis. Curr Opin Allergy Clin Immunol 12(6):616–622
Ragab D, Salah Eldin H, Taeimah M, Khattab R, Salem R (2020) The COVID-19 cytokine storm; what we know so far. Front Immunol 11:1446
Rahman SA, Adjeroh D (2015) Surface-based body shape index and its relationship with all-cause mortality. PLoS One 10(12):e0144639
Rajasingham R, Smith RM, Park BJ, Jarvis JN, Govender NP, Chiller TM, Denning DW, Loyse A, Boulware DR (2017) Global burden of disease of HIV-associated cryptococcal meningitis: an updated analysis. Lancet Infect Dis 17(8):873–881
Raviv Y, Kramer MR, Amital A, Rubinovitch B, Bishara J, Shitrit D (2007) Outbreak of aspergillosis infections among lung transplant recipients. Transpl Int 20(2):135–140
Riad A, Gomaa E, Hockova B, Klugar M (2021) Oral candidiasis of COVID-19 patients: case report and review of evidence. J Cosmet Dermatol 20(6):1580–1584
Riche CVW, Cassol R, Pasqualotto AC (2020) Is the frequency of candidemia increasing in COVID-19 patients receiving corticosteroids? J Fungi (Basel) 6(4):286
Rieber N, Gazendam RP, Freeman AF, Hsu AP, Collar AL et al (2016) Extrapulmonary *Aspergillus* infection in patients with CARD9 deficiency. JCI Insight 1(17):e89890
Rodrigues ML, Nosanchuk JD (2020) Fungal diseases as neglected pathogens: a wake-up call to public health officials. PLoS Negl Trop Dis 14(2):e0007964
Rodrigues CF, Rodrigues ME, Henriques M (2019) *Candida* sp. infections in patients with diabetes mellitus. J Clin Med 8(1):76
Romani L (2004) Immunity to fungal infections. Nat Rev Immunol 4(1):11–24
Roos D (2016) Chronic granulomatous disease. Br Med Bull 118(1):50–63
Rosebeck S, Madden L, Jin X, Gu S, Apel IJ et al (2011) Cleavage of NIK by the API2-MALT1 fusion oncoprotein leads to noncanonical NF-kappaB activation. Science 331(6016):468–472
Rudramurthy SM, Hoenigl M, Meis JF, Cornely OA, Muthu V, Gangneux JP, Perfect J, Chakrabarti A (2021) ECMM/ISHAM recommendations for clinical management of COVID-19 associated mucormycosis in low- and middle-income countries. Mycoses 64(9):1028–1037
Rueping MJ, Heinz WJ, Kindo AJ, Rickerts V, Lass-Floerl C et al (2010) Forty-one recent cases of invasive zygomycosis from a global clinical registry. J Antimicrob Chemother 65(2):296–302
Russell CD, Fairfield CJ, Drake TM, Turtle L, Seaton RA et al (2021) Co-infections, secondary infections, and antimicrobial use in patients hospitalised with COVID-19 during the first pandemic wave from the ISARIC WHO CCP-UK study: a multicentre, prospective cohort study. Lancet Microbe 2(8):e354–e365

Samaranayake LP, Fidel PL, Naglik JR, Sweet SP, Teanpaisan R, Coogan MM, Blignaut E, Wanzala P (2002) Fungal infections associated with HIV infection. Oral Dis 8(Suppl 2):151–160
Sánchez-Ramón S, Bermúdez A, González-Granado LI, Rodríguez-Gallego C, Sastre A, Soler-Palacín P (2019) Primary and secondary immunodeficiency diseases in oncohaematology: warning signs, diagnosis, and management. Front Immunol 10:586
Saud B, Bajgain P, Paudel G, Shrestha V, Bajracharya D et al (2020) Fungal infection among diabetic and Nnondiabetic individuals in Nepal. Interdiscip Perspect Infect Dis 2020:7949868
Savino W (2002) The thymus gland is a target in malnutrition. Eur J Clin Nutr 56(Suppl 3):S46–S49
Schaible UE, Kaufmann SH (2007) Malnutrition and infection: complex mechanisms and global impacts. PLoS Med 4(5):e115
Schmidt S, Hogardt M, Demir A, Röger F, Lehrnbecher T (2019) Immunosuppressive compounds affect the fungal growth and viability of defined *Aspergillus* species. Pathogens 8(4):273
Sefer AP, Abolhassani H, Ober F, Kayaoglu B, Bilgic Eltan S et al (2022) Expanding the clinical and immunological phenotypes and natural history of MALT1 deficiency. J Clin Immunol 42(3):634–652
Shirmast et al (2020) https://doi.org/10.22034/igj.2020.245559.1047
Shoham S, Marr KA (2012) Invasive fungal infections in solid organ transplant recipients. Future Microbiol 7(5):639–655
Silveira FP, Husain S, Kwak EJ, Linden PK, Marcos A et al (2007) Cryptococcosis in liver and kidney transplant recipients receiving anti-thymocyte globulin or alemtuzumab. Transpl Infect Dis 9(1):22–27
Singh R, Mullin GE (2017) A wasting syndrome and malnutrition caused by small intestine fungal overgrowth: case report and review of the literature. Integr Med (Encinitas) 16(3):48–51
Singh N, Alexander BD, Lortholary O, Dromer F, Gupta KL et al (2007) *Cryptococcus neoformans* in organ transplant recipients: impact of calcineurin-inhibitor agents on mortality. J Infect Dis 195(5):756–764
Singh AK, Singh R, Joshi SR, Misra A (2021) Mucormycosis in COVID-19: a systematic review of cases reported worldwide and in India. Diabetes Metab Syndr 15(4):102146
Sonoda M, Ishimura M, Eguchi K, Yada Y, Lenhartová N et al (2021) Progressive B cell depletion in human MALT1 deficiency. Clin Exp Immunol 206(3):237–247
Steinbach et al (2012) https://doi.org/10.1016/j.jinf.2012.08.003
Su HC, Jing H, Zhang Q (2011) DOCK8 deficiency. Ann N Y Acad Sci 1246(1):26–33
Sun Y, Huang H, Chen J, Li J, Ma J et al (2015) Invasive fungal infection in patients receiving chemotherapy for hematological malignancy: a multicenter, prospective, observational study in China. Tumour Biol 36(2):757–767
Sung JM, Ko WC, Huang JJ (2001) Candidaemia in patients with dialysis-dependent acute renal failure: aetiology, predisposing and prognostic factors. Nephrol Dial Transplant 16(12):2348–2356
Tissot F, Agrawal S, Pagano L, Petrikkos G, Groll AH et al (2017) ECIL-6 guidelines for the treatment of invasive candidiasis, aspergillosis and mucormycosis in leukemia and hematopoietic stem cell transplant patients. Haematologica 102(3):433–444
Tramsen L, Schmidt S, Roeger F, Schubert R, Salzmann-Manrique E et al (2014) Immunosuppressive compounds exhibit particular effects on functional properties of human anti-*Aspergillus* Th1 cells. Infect Immun 82(6):2649–2656
Tsai CS, Lee SS, Chen WC, Tseng CH, Lee NY et al (2022) COVID-19-associated candidiasis and the emerging concern of *Candida auris* infections. J Microbiol Immunol Infect 22:S1684-1182(1622)00283-00283
Tuano KS, Seth N, Chinen J (2021) Secondary immunodeficiencies: an overview. Ann Allergy Asthma Immunol 127(6):617–626
van Hal SJ, Marriott DJ, Chen SC, Nguyen Q, Sorrell TC et al (2009) Candidemia following solid organ transplantation in the era of antifungal prophylaxis: the Australian experience. Transpl Infect Dis 11(2):122–127

Vazquez JA (2010) Optimal management of oropharyngeal and esophageal candidiasis in patients living with HIV infection. HIV AIDS (Auckl) 2:89–101

Vento S, Cainelli F (2003) Infections in patients with cancer undergoing chemotherapy: aetiology, prevention, and treatment. Lancet Oncol 4(10):595–604

Vinh DC (2011) Insights into human antifungal immunity from primary immunodeficiencies. Lancet Infect Dis 11(10):780–792

Vinh DC, Patel SY, Uzel G, Anderson VL, Freeman AF et al (2010) Autosomal dominant and sporadic monocytopenia with susceptibility to mycobacteria, fungi, papillomaviruses, and myelodysplasia. Blood 115(8):1519–1529

Walsh TJ, Anaissie EJ, Denning DW, Herbrecht R, Kontoyiannis DP et al (2008) Treatment of aspergillosis: clinical practice guidelines of the Infectious Diseases Society of America. Clin Infect Dis 46(3):327–360

Wang Y, Zhang D, Hou Y, Shen S, Wang T (2020) The adaptor protein CARD9, from fungal immunity to tumorigenesis. Am J Cancer Res 10(8):2203–2225

Warnock DW (1998) Fungal infections in neutropenia: current problems and chemotherapeutic control. J Antimicrob Chemother 41(suppl_4):95–105

Williamson PR, Kwon-Chung KJ, Gallin JI (1992) Successful treatment of *Paecilomyces varioti* infection in a patient with chronic granulomatous disease and a review of *Paecilomyces* species infections. Clin Infect Dis 14(5):1023–1026

Winkelstein JA, Marino MC, Johnston RB Jr, Boyle J, Curnutte J et al (2000) Chronic granulomatous disease. Report on a national registry of 368 patients. Medicine (Baltimore) 79(3):155–169

Winkelstein JA, Marino MC, Ochs H, Fuleihan R, Scholl PR, Geha R, Stiehm ER, Conley ME (2003) The X-linked hyper-IgM syndrome: clinical and immunologic features of 79 patients. Medicine (Baltimore) 82(6):373–384

Yu H-H, Yang Y-H, Chiang B-L (2021) Chronic granulomatous disease: a comprehensive review. Clin Rev Allerg Immunol 61(2):101–113

Yusuf E, Seghers L, Hoek RAS, van den Akker JPC, Bode LGM, Rijnders BJA (2021) *Aspergillus* in critically ill COVID-19 patients: a scoping review. J Clin Med 10(11):2469

Zerbe & Holland (2005) https://doi.org/10.1086/432120

Zhang Q, Davis JC, Lamborn IT, Freeman AF, Jing H et al (2009) Combined immunodeficiency associated with DOCK8 mutations. N Engl J Med 361(21):2046–2055

# Epidemiology and Mechanisms of Antifungal Resistance in Common Fungal Infections

**Meher Rizvi, Nazish Fatima, and Hiba Sami**

## Introduction

Over the last century, the discovery and large-scale use of the antibiotics and antifungals have greatly advanced human and animal health. It truly appeared in the heady years of the 1950s and 1960s that the large-scale effect by the microbes on human health would be a thing of the past. However, less than a hundred years of the path breaking gains in healthcare, agriculture, and animal husbandry, the microbes have again tilted the scales and are fighting back by acquiring widespread resistance to antibiotics and antifungals currently in use.

It has been estimated that 13 million infections and 1.5 to 2 million deaths are attributable to fungal infections every year worldwide (Bongomin et al. 2017; McDermott 2022). With the rapidly increasing population of immunocompromised individuals, fungal infections are on the rise and can quickly progress from mild to moderate and severe form. About 1 in 50 people globally are immunocompromised due to a primary disease or as a consequence of immunosuppressive medication often with fatal outcomes (Corey et al. 2021). The increased pool is due to the rise of diabetes, malignancies, autoimmune disorders, stem cell and solid organ transplants, and rheumatoid arthritis in the population (Hoenigl et al. 2022).The advent of effective anti-retroviral medicines has, however, led to a reduction of fungal infections in HIV positive individuals (Wang et al. 2017). Viral infections, such as COVID-19 and influenza viruses, also predispose to fungal infections, such as Aspergillosis and Mucormycosis (Hoenigl et al. 2022).

M. Rizvi (✉)
Department of Microbiology and Immunology, College of Medicine and Health Sciences, Sultan Qaboos University, Muscat, Oman

N. Fatima · H. Sami
Department of Microbiology, Jawaharlal Nehru Medical College, Aligarh Muslim University, Aligarh, India



A. Gomes Rodrigues, A. Gupta (eds.), *Fungal Infections*, Fungal Biology,
https://doi.org/10.1007/978-3-032-06014-3_3

These are just a few of the clinical conditions that are significant drivers of fungal infections. Climate changes like droughts, flash floods, cyclones and tornadoes, human activities leading to expansion in urbanization, deforestation and destruction of native vegetation to expand agricultural landfall play a significant role in increasing the incidence of fungal infections (Wu et al. 2016; El-Sayed and Kamel 2020). Disaster-related fungi spread over extremely large area and cause infections in the lungs and soft tissues by types of fungi that are not usually found (Nnadi and Carter 2021). In 1977, a strong dust storm occurred in the southern San Joaquin Valley of California. This storm carried *Coccidioides immitis,* a dimorphic fungus, from Bakersfield (an area with a high occurrence of the fungus) to Sacramento County (where the fungus is uncommon). As a result, over 100 people were infected (Schneider et al. 1997).

Travel and tourism, as well as mass emigrations due to wars and famine, contribute to the spread of such infections as well (Vignier and Bouchaud 2018). Despite the changing demography and patient profile, the etiology of common fungal infections remains largely unchanged. *Aspergillus, Candida,* dimorphic fungi, such as *Histoplasma capsulatum, Blastomyces* spp., *Coccidioides* spp., *Mucormycetes, Cryptococcus* species, and *Pneumocystis jirovecii* cause the majority of fungal infections (Perlroth et al. 2007; Lee and Lau 2017). They have the potential to cause severe infections, leading to greater morbidity and mortality. *Candida albicans* is the main agent responsible for mucosal disease (Talapko et al. 2021), *Aspergillus fumigatus* for most allergic fungal disease (Sisodia and Bajaj 2024) and *Trichophyton* spp., especially *T. rubrum*, for skin infections (Blutfield et al. 2015).

Not only is the number of fungal infections increasing, but fungi are also becoming increasingly more challenging to treat. Traditionally, antifungal agents have been extensively used prophylactically and empirically in patients without the support of fungal diagnostic facilities in many parts of the world (Cowen et al. 2014). This continues to be very often the case in countries where such facilities may still be nonexistent or in their infancy. Even if present, the poor sensitivity of the diagnostic tools and the extended turnaround time of conventional culture-based techniques make evidence-based prescription of antifungals very difficult, if not impossible. In severely ill or vulnerable patients, delays in treatment can be potentially fatal. Thus, broad spectrum combination therapy has become the norm in such cases (Arendrup et al. 2013). The consequences of this are an increase in secondary resistance and breakthrough infections by resistant species. The outcome is that pathogenic yeasts and molds are rapidly acquiring resistance to the existing antifungal agents. The US CDC included *C. auris* as an "urgent threat" and *Aspergillus fumigatus* on the "watch list" of antibiotic-resistant threats (Vallabhaneni et al. 2017; Centers for Disease Control and Prevention (U.S.) 2019). This escalating resistance mirrors the exponential rise in antibiotic resistance. Unfortunately, therapeutic antifungal options are limited and increasing resistance means further shrinking of available therapeutic agents. The immensity of this problem is slowly gaining political and public attention. As the One Health approach gathers momentum, there is greater awareness about the close interlinking of climate, agriculture, environment, animal, and human health (Mackenzie and Jeggo 2019). The World Health Organization (WHO) has also emphasized that under One Health the well-being of humans, animals, and ecosystems is intricately interconnected. Alterations in these associations can escalate the likelihood of emergence and dissemination of novel diseases in both humans and animals (One Health 2024). FAO and UNEP have likely highlighted the importance of implementing strategies to mitigate antifungal resistance. This could include promoting responsible use of antifungal agents in agriculture and veterinary medicine, encouraging surveillance and monitoring of antifungal resistance in various settings, supporting research into new antifungal treatments, and raising awareness among stakeholders about the risks associated with antifungal resistance. Both organizations are likely advocating for a One Health approach, which recognizes the interconnectedness of human health, animal health, and the environment in addressing issues like antimicrobial resistance, including antifungal resistance.

Fungal infections are a One Health issue—the environment, plants, and animals serve as a reservoir for these infections. Thus, fungal infections can spread from all these sources to humans. Antifungals from the same families are used in agriculture, animal husbandry, veterinaries, and hospitals. In fact antifungals are used far more widely for protecting plants in agriculture than in treating humans (Toda et al. 2021).With greater and closer collaboration between the four major antifungal consuming sectors, it is becoming clear that antifungal resistance in humans is being spearheaded by exposure of the human fungal pathogens to antifungals being used in agriculture. There is a very real fear of the spread of resistance from the environment to humans. Thus, to find a sustainable solution for preserving antifungals, we will have to adopt a multipronged One Health approach.

In this chapter, we will discuss the prevalence of antifungal resistance to common fungal infections, mechanism of action of the antifungal agents, the mechanism of antifungal resistance as well as explore alternative treatment strategies.

## Antifungal Drugs and Their Mechanism of Action

In contrast to thousands of antibiotics available to combat bacterial infections, the number of antifungal agents is indeed limited. There are around ten classes of antifungal drugs available for clinical use (Fisher et al. 2018; Perfect 2017). This is because since both humans and fungi are eukaryotic, the number of effective targets against fungi are limited, compared to bacteria, which are prokaryotes. Currently there are five groups of pharmaceuticals and with mounting resistance, effective antifungal resources are shrinking. Box 1 illustrates the common antifungal agents and their mechanisms of action.

**Box 1 Common Antifungal Agents and Their Mechanisms of Action**

1. Griseofulvin inhibits mitosis and acts on dermatophytes.
2. Imidazoles, such as clotrimazole, miconazole, ketoconazole, inhibit lanosterol synthesis.
3. Triazoles, such as fluconazole, itraconazole, voriconazole, posaconazole, isavuconazole, inhibit ergosterol synthesis.
4. Echinocandins (e.g., caspofungin, micafungin, anidulafungin) inhibit the cell wall formation.
5. Polyenes (e.g., amphotericin B) induces membrane pore formation.
6. Flucytocine inhibits nucleic acid synthesis.

## Antifungal Resistance

Antifungal resistance, intrinsic or acquired, is developing at an alarming pace and has grave consequences. *Cryptococcus* is intrinsically resistant to echinocandins while *Aspergillus fumigatus* is acquiring resistance to the azole group. One of the first reports of azole resistance was by Ryley et al. in *Candida albicans* in chronic mucocutaneous candidiasis in 1980 (Ryley et al. 1984). Dermatophyte resistance to griseofulvin was reported even earlier, in 1961 (Michaelides 1961). Azole resistance in dermatophytes is 19% in some parts of the world, e.g., India (Ghannoum 2016). A study published in the Indian Journal of Dermatology in 2018 reported azole resistance rates of up to 19.5% among clinical isolates of dermatophytes in India. The study analyzed data from multiple centers across India and found varying levels of resistance among different species of dermatophytes (Mahajan et al. 2017). Another study published in the Journal of Clinical and Diagnostic Research in 2016 evaluated the antifungal susceptibility patterns of dermatophyte isolates from patients with superficial dermatophytosis in South India. The study found that 18.4% of the isolates exhibited resistance to azoles (Dogra and Uprety 2016). In 2003 *Trichophyton rubrum* isolated from a patient with onychomycosis displayed intrinsic resistance to terbinafine (Mukherjee et al. 2003). The rapidly emerging yeast *Candida auris* is not only multidrug resistant but has shown tremendous potential to spread in the hospital setting (Vallabhaneni et al. 2017).

Fungal resistance can be mediated by several factors. Fungal factors that mediate resistance are (i) increased efflux of the medicines, (ii) reduced affinity for the therapeutic agent, (iii) alterations in cell response, (iv) altered metabolism, and (v) biofilm formation, leading to development of resistance to antifungals (Revie et al. 2018). Resistance is caused by many molecular mechanisms. Point mutations, gene deletion or amplification, gene transfer, loss of regulatory elements, and/or transcriptional activation being some of them (Lee et al. 2023). Immunosuppression in hosts due to HIV, solid organ transplant, human stem cell transplant, and diabetes plays a significant role in development of resistance. Delay in initiation of appropriate therapy, inappropriate doses, poor API absorption to the site of infection, such as abscesses or nails, insufficient knowledge of pharmacokinetics, interactions of antifungals, and lack of fungicidal agents play a role in building resistance (Revie et al. 2018). In the following two sections, resistance in Candidiasis and Aspergillosis will be discussed in furb

## Candidiasis

Candida infections are becoming increasingly difficult to treat due to emergence of resistance to antifungals (Ksiezopolska and Gabaldón 2018). *Candida albicans* is the most common cause of Candida infections across all spectra of disease, ranging from mild to severe. It is responsible for the majority of mucosal and invasive cases of candidiasis. Although it is intrinsically sensitive to antifungals, prolonged or repeated exposures to antifungals have resulted in escalation of resistance.

Antifungal resistance is most pronounced in the non-albicans Candida, such as *Candida auris, Candida glabrata, Candida parapsilosis, C. krusei* and *C. tropicalis* (Ksiezopolska and Gabaldón 2018). The increasing resistance to fluconazole is worrisome as it is not only the most commonly used but also the most easily available antifungal agent around the globe. It is useful for both prophylaxis and treatment of Candida infections (Arendrup et al. 2013). Another cause of concern is that the non-*Candida albicans* species have a greater potential to cause outbreaks (Bongomin et al. 2017). According to CDC, fluconazole resistance was observed in 7% of *Candida* sp. isolated from blood stream infections (Toda et al. 2019). Fluconazole resistance in Candida species can arise from various mechanisms, including alterations in the target enzyme (lanosterol 14α-demethylase encoded by the *ERG11* gene), efflux pump overexpression, and changes in membrane permeability. These mechanisms can occur individually or in combination, leading to decreased susceptibility to fluconazole (Sanglard and Coste 2016). Fluconazole resistance in Candida species has important clinical implications, as it can lead to treatment failure and the need for alternative antifungal agents, which may have higher toxicity or limited availability. Additionally, fluconazole-resistant Candida species are often associated with healthcare-associated infections and outbreaks in hospital settings, posing challenges for infection control and patient management (Guinea 2014).

*C. glabrata* has emerged as a significant cause of mucosal and bloodstream infections (Ksiezopolska and Gabaldón 2018). The gastrointestinal tract is considered an important site for selection of resistant *C. glabrata.* It comes equipped with intrinsic hetero-resistance to azoles and is in the process of evolving stable resistance to both azoles and echinocandins (Ksiezopolska and Gabaldón 2018). Increased medicine exposure is driving this resistance as well. As the pool of individuals susceptible to such infections rises worldwide, multidrugresistant *C. glabrata* are also expected to rise (Healey and Perlin 2018). A worldwide increase in prevalence of fluconazole-resistant *C. glabrata* has been observed by the SENTRY

Antifungal Surveillance Programme (Pfaller et al. 2019). They have reported that resistance increased from 8.6% in 1997 to 10.1% in 2014. Echinocandin resistance in the same report across the globe ranged from 1.7–3.5%. What is concerning is that co-carriage of resistance to echinocandins and azoles was reported among 5.5–7.6% *C. glabrata* isolates (Pfaller et al. 2019).

Echinocandins have a favorable safety profile and broad-spectrum anti-*Candida* activity (Ksiezopolska and Gabaldón 2018). They target GS *FKS* subunits and inhibit β-(1,3)-D-glucan synthase. These subunits are encoded by *FKS1*, *FKS2*, and *FKS3*. They are the antifungals of choice in invasive candidiasis because of their broad-spectrum of activity. The resistance is mediated by amino acid substitutions in *FKS1 C. albicans*, *C. tropicalis*, *C. krusei*, *C. glabrata*, while in *C. glabrata* it is also due to substitutions in *FKS2*. The escalation of resistance to echinocandins is concerning (Alexander et al. 2013) and could significantly narrow down the available treatment options. Liposomal amphotericin B, which is more toxic, becomes the agent of choice in such isolates (Ksiezopolska and Gabaldón 2018). As can be expected, mortality is higher in patients infected with drug-resistant Candida (Ksiezopolska and Gabaldón 2018).

While the world faces the increasing problem of antifungal resistance, new and more dangerous species are emerging. *C. auris* is one such recently emerged fungal pathogen, which was first described in 2009 (Vallabhaneni et al. 2017). Its emergence and association with invasive disease was reported concomitantly from six continents. It poses a grave public health threat and is considered an urgent global threat (Vallabhaneni et al. 2017). It can spread efficiently in healthcare facilities, causing prolonged healthcare-associated outbreaks. This pathogen has a unique ability to colonize the human skin and mucosa, and persists on surfaces in hospital and nursing home environments, including bed rails and medical equipment, causing difficult-to-eradicate outbreaks (Vallabhaneni et al. 2017).

It is also very difficult to eliminate it. *C. auris* exhibits a far greater resistance to a wider spectrum of antifungals than other Candida species. It is resistant to azoles, polyenes, and sometimes echinocandins (Lockhart et al. 2017). To complicate the matter further *C. auris* is difficult to identify in routine microbiology laboratories, making appropriate management more challenging. *C. auris* has been reported as the leading cause of candidemia in many parts of the world with prevalence as high as 40% being reported in India, 38% in Kenya, and 14% in South Africa (Shastri et al. 2020; Adam et al. 2019; Gow et al. 2022). It has been reported that over 90% of *C. auris* isolates are resistant to fluconazole and a third are resistant to amphotericin B (Pfaller et al. 2019). They may appear susceptible to echinocandins but develop resistance during treatment. In a multicentric study, spanning three continents, fluconazole resistance was observed in 93% of isolates, while amphotericin B and echinocandin resistance was observed in 35% and 7% isolates, respectively (Lockhart et al. 2017). *Candida* species *haemulonii, krusei,* and *lusitaniae,* which are the closest related species to *C. auris*, also manifest high resistance to fluconazole, amphotericin B, or both (Lockhart et al. 2017).

On whole genome sequence analysis, high frequency mutations have been reported in the drug targets, *ERG11*, *FKS1*, *Tac1B*, and *Mrr1* transcription factors (van Schalkwyk et al. 2019). These regulate the expression of drug transporters that increase the resistance to azoles and echinocandins. Utilizing population genomic analysis, there may be a difference in both the nature and the frequency of these mutations among the four major genetic clades of *C. auris* (Chow et al. 2020).

Amphotericin resistance may be linked to mutations leading to loss of function in the *ERG6* protein, which is involved in ergosterol biosynthesis (Adam et al. 2019). Prior antifungal treatment, older age, increased length of stay in the hospital or ICU, and central venous catheters play a role in developing *C. auris* candidemia (Shor and Perlin 2015).

## Aspergillosis

The genus *Aspergillus* consists of a wide variety of species that can thrive and reproduce within several environmental conditions including indoor and outdoor environments, especially in organic debris and soil (Mousavi et al. 2016).There are more than a hundred species of Aspergillus but only a few have been implicated in human disease. Although 90% of human infections are caused by *Aspergillus fumigatus* (Mousavi et al. 2016), other Aspergilli such as *Aspergillus flavus, Aspergillus terreus, Aspergillus niger, Aspergillus lenustus,* and *Aspergillus alliaceus* have been identified as pathogens less frequently. Resistance of Aspergillus species to antifungal drugs, including azoles, echinocandins, and polyenes, has emerged as a grave public health crisis in recent years. The clinical implications of antifungal resistance in Aspergillus species are significant. Patients with invasive aspergillosis who are infected with resistant strains have higher mortality rates and may require more aggressive treatment strategies (Latgé and Chamilos 2019).

Aspergillus species are associated with serious diseases ranging from hypersensitivity disorders to rapidly invasive disseminated diseases. Allergic bronchopulmonary aspergillosis, chronic pulmonary aspergillosis, aspergilloma, cutaneous aspergillosis, and invasive aspergillosis are the most common infections associated with Aspergillus. Over three million people worldwide are living with chronic pulmonary aspergillosis (CPA). CPA is a broad term that includes simple aspergilloma, chronic cavitary pulmonary aspergillosis, and chronic fibrosing pulmonary aspergillosis. It is an uncommon pulmonary disease often causing infection in individuals who have underlying respiratory diseases like chronic obstructive pulmonary disease (COPD), tuberculosis or sarcoidosis (Denning et al. 2016).

Allergic bronchopulmonary aspergillosis (ABPA), a progressive allergic lung disease, commonly aggravates asthma and cystic fibrosis. It may manifest as poorly controlled asthma, hemoptysis, fever, and expectoration of mucus plugs. *Aspergillus fumigatus* is a common airborne fungus responsible for invasive pulmonary aspergillosis (IPA) in immunocompromised patients. In profoundly immunocompromised individuals, *A. fumigatus* presents as an invasive, aggressive aspergillosis, an aggressive disease that leads to angio-invasion, tissue destruction, and septic shock (Kosmidis and Denning 2015). The treatment of IPA mainly relies on triazole antifungals, such as voriconazole, itraconazole, and posaconazole, echinocandins, and polyenes, such as amphotericin B. However, the emergence of resistance to these antifungal agents in *A. fumigatus* has become a major concern in recent years (Latgé and Chamilos 2019).

The prevalence of antifungal resistance in CPA and ABPA varies worldwide. A study conducted by Chakrabarti in India found that 22.9% of *Aspergillus fumigatus* isolates were resistant to azoles, while another study from Brazil reported 11% azole-resistant Aspergillus infections. Spain reported 9.7% resistance to azoles, while the Netherlands reported 11.2% resistance (Chakrabarti et al. 2011; Escribano et al. 2021; Burks et al. 2021). This resistance is primarily due to mutations in the genes responsible for the synthesis of the target proteins or changes in the expression of these genes.

Triazole resistance in *Aspergillus* spp. has been reported from both environmental and clinical isolates (Rivero-Menendez et al. 2016). Widescale fungicide use in agriculture and in clinical practice tend to select for triazole resistance, which varies from 1–10%. A study from Netherlands reported more than 25% resistant isolates (Lestrade et al. 2019). However, this number may be an underestimate as rigorous surveillance is hindered due to two reasons: difficulty in obtaining appropriate respiratory specimens from the immunocompromised individuals and limited availability as well as high cost entailed in determining phenotypic (MIC) or genotypic (*CYP51A*) azole susceptibility testing in hospital laboratories. An azole resistance of 19% was reported from ICU patients in a Dutch study, with higher six-week mortality associated with such infections (Mortensen et al. 2015).

The increased incidence and treatment of concomitant aspergillosis with spike in influenza and COVID-19 infections too drove triazole resistance (Meijer et al. 2020). Azole resistance is primarily due to alterations in the target enzyme, lanosterol 14-alpha-demethylase (*CYP51A*), which is encoded by the *CYP51A* gene. Mutations in this gene result in reduced affinity of the enzyme for azole agents, leading to their decreased efficacy. Additionally, overexpression of the *CYP51A* gene can lead to increased resistance to azoles. Other mechanisms of azole resistance include increased API efflux and alterations in the sterol biosynthesis pathway (Rivero-Menendez et al. 2016).

Echinocandin resistance is linked to mutations in the *FKS* genes, which encode for the β-1,3-D-glucan synthase enzyme (Perlin 2015). Mutations in these genes result in reduced susceptibility to echinocandins, leading to treatment failure. Other mechanisms of echinocandin resistance include alterations in the expression of genes involved in cell wall biosynthesis and increased drug efflux.

Polyene resistance develops due to alterations in the ergosterol biosynthesis pathway (Cowen et al. 2014). Mutations in genes encoding enzymes involved in ergosterol biosynthesis, such as *ERG11*, result in decreased susceptibility to polyenes. Additionally, alterations in the expression of genes involved in ergosterol biosynthesis and increased drug efflux can also lead to polyene resistance. Voriconazole-echinocandin combination therapy or liposomal amphotericin B are being recommended by European guidelines where resistance to triazoles are higher than 10% (Ullmann et al. 2018).

The emergence of resistance to azoles, echinocandins, and polyenes in *A. fumigatus* is a significant clinical challenge. Understanding of the molecular mechanisms of resistance is critical to developing effective treatment strategies and preventing the spread of resistant strains. There is an urgent need for more research to understand the epidemiology, diagnosis, and management of azole-resistant *Aspergillus fumigatus*.

## Conclusion

A One Health approach, a robust antifungal stewardship, institution of effective infection prevention, control measures, and availability of good fungal diagnostic laboratories all over the world can ensure appropriate patient management and reduce or stave off antifungal resistance. It is important to develop alternative antifungal agents for agricultural purposes, which do not cross-react with those needed to treat humans. This would reduce the pressure on existing antifungals. The availability of easily accessible, sensitive, and inexpensive diagnostic tools, such as specific biomarkers or molecular assays, is essential for detecting antifungal resistance and guiding evidence-based prescription of antifungal agents. Molecular mechanisms of resistance, such as mutations in target genes or alterations in drug efflux pumps, can be detected using these diagnostic tools, enabling tailored antifungal therapy and informing antifungal stewardship programs. This would enable the development of a strong antifungal stewardship program, which would reduce inappropriate antifungal prescriptions. All these efforts would help in reducing resistance without affecting clinical outcomes.

## References


Adam RD, Revathi G, Okinda N, Fontaine M, Shah J, Kagotho E et al (2019) Analysis of *Candida auris* fungemia at a single facility in Kenya. Int J Infect Dis 85:182–187

Alexander BD, Johnson MD, Pfeiffer CD, Jiménez-Ortigosa C, Catania J, Booker R et al (2013) Increasing echinocandin resistance in *Candida glabrata*: clinical failure correlates with presence of FKS mutations and elevated minimum inhibitory concentrations. Clin Infect Dis Off Publ Infect Dis Soc Am 56(12):1724–1732

Arendrup MC, Dzajic E, Jensen RH, Johansen HK, Kjaldgaard P, Knudsen JD et al (2013) Epidemiological changes with potential implication for antifungal prescription recommendations for fungaemia: data from a nationwide fungaemia surveillance programme. Clin Microbiol Infect 19(8):e343–e353

Blutfield MS, Lohre JM, Pawich DA, Vlahovic TC (2015) The immunologic response to *Trichophyton Rubrum* in lower extremity fungal infections. J Fungi. 1(2):130–137

Bongomin F, Gago S, Oladele RO, Denning DW (2017) Global and multi-national prevalence of fungal diseases-estimate precision. J Fungi. 1–29:57

Burks C, Darby A, Gómez Londoño L, Momany M, Brewer MT (2021) Azole-resistant *Aspergillus fumigatus* in the environment: identifying key reservoirs and hotspots of antifungal resistance. Xue C, editor. PLOS Pathog. 17(7):e1009711

Centers for Disease Control and Prevention (U.S.) (2019) Antibiotic resistance threats in the United States, 2019. Centers for Disease Control and Prevention (U.S.); [cited 2024 Mar 26]. Available from: https://stacks.cdc.gov/view/cdc/82532

Chakrabarti A, Chatterjee SS, Das A, Shivaprakash MR (2011) Invasive aspergillosis in developing countries. Med Mycol 49(S1):S35–S47

Chow NA, Muñoz JF, Gade L, Berkow EL, Li X, Welsh RM et al (2020) Tracing the evolutionary history and global expansion of *Candida auris* using population genomic analyses. Butler G, Nielsen K, editors. mBio 11(2):e03364–e03319

Corey L, Beyrer C, Cohen MS, Michael NL, Bedford T, Rolland M (2021) SARS-CoV-2 variants in patients with immunosuppression. N Engl J Med 385(6):562–566

Cowen LE, Sanglard D, Howard SJ, Rogers PD, Perlin DS (2014) Mechanisms of antifungal drug resistance. Cold Spring Harb Perspect Med. 5(7):a019752

Denning DW, Cadranel J, Beigelman-Aubry C, Ader F, Chakrabarti A, Blot S et al (2016) Chronic pulmonary aspergillosis: rationale and clinical guidelines for diagnosis and management. Eur Respir J 47(1):45–68

Dogra S, Uprety S (2016) The menace of chronic and recurrent dermatophytosis in India: is the problem deeper than we perceive? Indian Dermatol Online J 7(2):73–76

El-Sayed A, Kamel M (2020) Climatic changes and their role in emergence and re-emergence of diseases. Environ Sci Pollut Res Int 27(18):22336–22352

Escribano P, Rodríguez-Sánchez B, Díaz-García J, Martín-Gómez MT, Ibáñez-Martínez E, Rodríguez-Mayo M et al (2021) Azole resistance survey on clinical *Aspergillus fumigatus* isolates in Spain. Clin Microbiol Infect 27(8):1170.e1–1170.e7

Fisher MC, Hawkins NJ, Sanglard D, Gurr SJ (2018) Worldwide emergence of resistance to antifungal drugs challenges human health and food security. Science 360(6390):739–742

Ghannoum M (2016) Azole resistance in dermatophytes. J Am Podiatr Med Assoc 106(1):79–86

Gow NAR, Johnson C, Berman J, Coste AT, Cuomo CA, Perlin DS et al (2022) The importance of antimicrobial resistance in medical mycology. Nat Commun 13(1):5352

Guinea J (2014) Global trends in the distribution of Candida species causing candidemia. Clin Microbiol Infect Off Publ Eur Soc Clin Microbiol Infect Dis 20(Suppl 6):5–10

Healey KR, Perlin DS (2018) Fungal resistance to Echinocandins and the MDR phenomenon in *Candida glabrata*. J Fungi. 4(3):105

Hoenigl M, Seidel D, Sprute R, Cunha C, Oliverio M, Goldman GH et al (2022) COVID-19-associated fungal infections. Nat Microbiol 7(8):1127–1140

Kosmidis C, Denning DW (2015) The clinical spectrum of pulmonary aspergillosis. Thorax 70(3):270–277

Ksiezopolska E, Gabaldón T (2018) Evolutionary emergence of drug resistance in Candida opportunistic pathogens. Genes 9(9):461

Latgé JP, Chamilos G (2019) *Aspergillus fumigatus* and Aspergillosis in 2019. Clin Microbiol Rev 33(1):e00140–e00118

Lee PP, Lau YL (2017) Cellular and molecular defects underlying invasive fungal infections-revelations from endemic mycoses. Front Immunol 8:735

Lee Y, Robbins N, Cowen LE (2023) Molecular mechanisms governing antifungal drug resistance. Npj Antimicrob Resist 1(1):1–9

Lestrade PPA, Meis JF, Melchers WJG, Verweij PE (2019) Triazole resistance in *Aspergillus fumigatus*: recent insights and challenges for patient management. Clin Microbiol Infect 25(7):799–806

Lockhart SR, Etienne KA, Vallabhaneni S, Farooqi J, Chowdhary A, Govender NP et al (2017) Simultaneous emergence of multidrug-resistant *Candida auris* on 3 continents confirmed by whole-genome sequencing and epidemiological analyses. Clin Infect Dis 64(2):134–140

Mackenzie JS, Jeggo M (2019) The one health approach—why is it so important? Trop Med Infect Dis 4(2):88

Mahajan S, Tilak R, Kaushal SK, Mishra RN, Pandey SS (2017) Clinico-mycological study of dermatophytic infections and their sensitivity to antifungal drugs in a tertiary care center. Indian J Dermatol Venereol Leprol 83(4):436–440

McDermott A (2022) Drug-resistant fungi on the rise. Proc Natl Acad Sci 119(48):e2217948119

Meijer EFJ, Dofferhoff ASM, Hoiting O, Buil JB, Meis JF (2020) Azole-resistant COVID-19-associated pulmonary Aspergillosis in an immunocompetent host: a case report. J Fungi. 6(2):79

Michaelides P (1961) *Trichophyton Tonsurans* infection resistant to Griseofulvin: a case demonstrating clinical and in vitro resistance. Arch Dermatol 83(6):988

Mortensen KL, Denning DW, Arendrup MC (2015) The burden of fungal disease in Denmark. Mycoses 58:15–21

Mousavi B, Hedayati M, Hedayati N, Ilkit M, Syedmousavi S (2016) Aspergillus species in indoor environments and their possible occupational and public health hazards. Curr Med Mycol 2(1):36–42

Mukherjee PK, Leidich SD, Isham N, Leitner I, Ryder NS, Ghannoum MA (2003) Clinical *Trichophyton rubrum* strain exhibiting primary resistance to Terbinafine. Antimicrob Agents Chemother 47(1):82–86

Nnadi NE, Carter DA (2021) Climate change and the emergence of fungal pathogens. PLoS Pathog 17(4):e1009503

One Health [cited 2024 May 2]. Available from: https://www.who.int/news-room/fact-sheets/detail/one-health

Perfect JR (2017) The antifungal pipeline: a reality check. Nat Rev Drug Discov 16(9):603–616

Perlin DS (2015) Mechanisms of echinocandin antifungal drug resistance. Ann N Y Acad Sci 1354(1):1–11

Perlroth J, Choi B, Spellberg B (2007) Nosocomial fungal infections: epidemiology, diagnosis, and treatment. Med Mycol 45(4):321–346

Pfaller MA, Diekema DJ, Turnidge JD, Castanheira M, Jones RN (2019) Twenty years of the SENTRY antifungal surveillance program: results for Candida species from 1997–2016. Open Forum Infect Dis 6(Suppl 1):S79–S94

Revie NM, Iyer KR, Robbins N, Cowen LE (2018) Antifungal drug resistance: evolution, mechanisms and impact. Curr Opin Microbiol 45:70–76

Rivero-Menendez O, Alastruey-Izquierdo A, Mellado E, Cuenca-Estrella M (2016) Triazole resistance in *Aspergillus spp.*: a worldwide problem? J Fungi 2(3):21

Ryley JF, Wilson RG, Barrett-Bee KJ (1984) Azole resistance in *Candida albicans*. Sabouraudia 22(1):53–63

Sanglard D, Coste AT (2016) Activity of Isavuconazole and other azoles against Candida clinical isolates and yeast model systems with known azole resistance mechanisms. Antimicrob Agents Chemother 60(1):229–238

Schneider E, Hajjeh RA, Spiegel RA, Jibson RW, Harp EL, Marshall GA et al (1997) A coccidioidomycosis outbreak following the Northridge, Calif, earthquake. JAMA 277(11):904–908

Shastri PS, Shankarnarayan SA, Oberoi J, Rudramurthy SM, Wattal C, Chakrabarti A (2020) *Candida auris* candidaemia in an intensive care unit—prospective observational study to evaluate epidemiology, risk factors, and outcome. J Crit Care 57:42–48

Shor E, Perlin DS (2015) Coping with stress and the emergence of multidrug resistance in fungi. PLoS Pathog 11(3):e1004668

Sisodia J, Bajaj T (2024) Allergic Bronchopulmonary Aspergillosis. In: StatPearls. StatPearls Publishing, Treasure Island (FL) [cited 2024 Mar 26]. Available from: http://www.ncbi.nlm.nih.gov/books/NBK542329/

Talapko J, Juzbašić M, Matijević T, Pustijanac E, Bekić S, Kotris I et al (2021) *Candida albicans*-the virulence factors and clinical manifestations of infection. J Fungi Basel Switz 7(2):79

Toda M, Williams SR, Berkow EL, Farley MM, Harrison LH, Bonner L et al (2019) Population-based active surveillance for culture-confirmed Candidemia — four sites, United States, 2012–2016. MMWR Surveill Summ 68(8):1–15

Toda M, Beer KD, Kuivila KM, Chiller TM, Jackson BR (2021) Trends in agricultural Triazole fungicide use in the United States, 1992–2016 and possible implications for antifungal-resistant fungi in human disease. Environ Health Perspect 129(5):EHP7484, 055001

Ullmann AJ, Aguado JM, Arikan-Akdagli S, Denning DW, Groll AH, Lagrou K et al (2018) Diagnosis and management of Aspergillus diseases: executive summary of the 2017 ESCMID-ECMM-ERS guideline. Clin Microbiol Infect 24:e1–e38

Vallabhaneni S, Kallen A, Tsay S, Chow N, Welsh R, Kerins J et al (2017) Investigation of the first seven reported cases of *Candida auris*, a globally emerging invasive, multidrug-resistant fungus-United States, May 2013-August 2016. Am J Transplant Off J Am Soc Transplant Am Soc Transpl Surg 17(1):296–299

van Schalkwyk E, Mpembe RS, Thomas J, Shuping L, Ismail H, Lowman W et al (2019) Epidemiologic shift in Candidemia driven by *Candida auris*, South Africa, 2016–20171. Emerg Infect Dis 25(9):1698–1707

Vignier N, Bouchaud O (2018) Travel, migration and emerging infectious diseases. EJIFCC 29(3):175–179

Wang RJ, Miller RF, Huang L (2017) Approach to fungal infections in HIV-infected individuals: pneumocystis and beyond. Clin Chest Med 38(3):465–477

Wu X, Lu Y, Zhou S, Chen L, Xu B (2016) Impact of climate change on human infectious diseases: empirical evidence and human adaptation. Environ Int 86:14–23

# Zearalenone Production: Occurrence, Biosynthesis in *Fusarium* spp., and Impacts on Public Health

**Gülruh Albayrak, Tuğba Teker, Gülin İnci Varol, and Emre Yörük**

## Mycotoxins and Mycotoxicoses

Mycotoxins are the products of fungal secondary metabolism. They can be classified regarding their chemical structures (e.g., lactones, coumarins), biosynthetic origins (polyketides, amino acid-derived, etc.), and myriad biological effects. Cell biologists have grouped them as teratogens, mutagens, carcinogens, and allergens more generically. Clinicians classify mycotoxins as hepatotoxins, nephrotoxins, neurotoxins, and immunotoxins according to the affected organ (Bennett and Klich 2003). Although mycotoxins are produced by more than 100 mold species, *Aspergillus*, *Penicillium*, and *Fusarium* spp. are the main producers (Mahato et al. 2021). Some of mycotoxigenic fungi penetrate the crops from their spike, flower, seed, grain, root, etc., before harvest, while others invade crops during storage. Therefore, they are known as field fungi and storage fungi. Field fungi are more prevalent when the moisture content is high (18%–30%). However, storage fungi generally require dry conditions (14%–16% moisture content) for growing well (Gurikar et al. 2023).

G. Albayrak (✉) · G. İ. Varol
Science Faculty, Department of Molecular Biology and Genetics, Istanbul University, Vezneciler, Istanbul, Türkiye
e-mail: gulruh@istanbul.edu.tr; gulin.varol@istanbul.edu.tr

T. Teker
Faculty of Engineering and Natural Sciences, Department of Molecular Biology and Genetics, Istanbul Atlas University, Kağıthane, Istanbul, Türkiye
e-mail: tugba.teker@atlas.edu.tr

E. Yörük
Faculty of Arts and Sciences, Department of Molecular Biology and Genetics, Istanbul Yeni Yuzyil University, Cevizlibağ, Istanbul, Türkiye
e-mail: emre.yoruk@yeniyuzyil.edu.tr



A. Gomes Rodrigues, A. Gupta (eds.), *Fungal Infections*, Fungal Biology,
https://doi.org/10.1007/978-3-032-06014-3_4

Mycotoxin contamination is a major public health and economic concern (Janik et al. 2020). People are exposed to mycotoxins through dermal route, inhalation, and by consuming contaminated food. Contaminated foods are the most common causes of acute and chronic toxic diseases called mycotoxicoses (Murray et al. 2020). Aflatoxins, ochratoxin A, patulin, trichothecenes, fumonisins, and zearalenone (ZEN) are the most common mycotoxins identified in food so far (Janik et al. 2020). A variety of illnesses and clinical syndromes depending on exposition of mycotoxin in humans were reported (Table 1). Most fungi are able to grow at human body temperature. However, the optimum temperature for the biosynthesis of most mycotoxins is lower (20 °C to 30 °C) than human body temperature (37 °C). Therefore, the exact baseline about mycotoxin exposure during the course of fungal infections is still unknown (Murray et al. 2020).

Since it is undeniable that contaminated food poses a significant risk to human health, food safety is important in maintaining human health and well-being. In this context, the Joint Food and Agriculture Organization of the United Nations (FAO)/

**Table 1** Mycotoxin-related illnesses and their symptoms

| Toxin | Fungus | Diseases: Clinical Symptoms |
|---|---|---|
| Aflatoxins | *Aspergillus* spp. | *Acute aflatoxicosis*: Acute hepatitis with centrilobular necrosis and steatosis by histopathologic examination (Smith and McGinnis 2009)<br>*Kwashiorkor*: Protein deficiency, hepatic steatosis, and ascites (Smith and McGinnis 2009)<br>*Liver cancer* (Smith and McGinnis 2009; Janik et al. 2020) |
| Ochratoxin | *Aspergillus* spp.<br>*Penicillium* spp. | *Chronic nephritis (Balkan nephropathy)*: Chronic renal injury (Smith and McGinnis 2009; Peraica 2016) |
| Trichothecenes | *Fusarium* spp.<br>*Trichoderma* spp.<br>*Myrothecium* spp.<br>*Phomopsis* spp.<br>*Trichothecium* spp.<br>*Stachybotrys* spp. | *Alimentary toxic aleukia*: Oral mucosal ulcerations and gastroenteritis were followed by pancytopenia due to bone marrow toxicity, with hemorrhage and agranulocytosis (Smith and McGinnis 2009)<br>*Red-mold toxicosis*: Nausea, vomiting, abdominal pain, diarrhea, chills and headache (Kuhn and Ghannoum 2003)<br>*Toxic mould syndrome/sick building syndrome/building-related illness*: Fatigue, headache, irritation of the eyes, nose and pharynx, a variety of neurologic symptoms (Kuhn and Ghannoum 2003; Smith and McGinnis 2009) |
| Fumonisins | *Fusarium* spp. | *Esophageal cancer*: Dysphagia, pain, hemorrhage (Murray et al. 2020)<br>*Equine leukomalacia*: High incidence of anencephaly and other neural tube defects along (Smith and McGinnis 2009) |
| Zearalenone | *Fusarium* spp. | Alterations in female reproductive organs and associated health problems (Peraica 2016; Rogowska et al. 2019; Mahato et al. 2021)<br>*Scabby grain toxicoses*: Nausea, vomiting, and diarrhea (Zain 2011) |

**Table 2** Institutions that carry out scientific assessments for food and feed safety to safeguard public health

| Institution | Web Site (https://) |
|---|---|
| Agricultural and Processed Food Products Export Development Authority (APEDA) | apeda.gov.in/apedawebsite/ |
| European Commission | food.ec.europa.eu/index_en |
| European Food Safety Authority (EFSA) | www.efsa.europa.eu/en |
| Food Standards Agency (FSA) | www.food.gov.uk/ |
| Food Safety and Standard Authority of India (FSSAI) | www.fssai.gov.in/ |
| German Federal Institute for Risk Assessment (BfR) | www.bfr.bund.de/en/home.html |
| Republic of Türkiye Ministry of Agriculture and Forestry | https://www.tarimorman.gov.tr/Sayfalar/EN/AnaSayfa.aspx |
| State General Laboratory (SGL) | www.moh.gov.cy/moh/sgl/sgl.nsf/home_en/home_en?opendocument |
| Tokyo Food Safety Information Center | www.fukushihoken.metro.tokyo.lg.jp/shokuhin/eng/anzenjoho_index.html |
| U.S. Food &Drug Administration (FDA) | www.fda.gov/ |
| World Health Organization (WHO) | www.who.int/ |

World Health Organization (WHO) Expert Committee on Food Additives (JECFA), which is an international scientific committee, has been evaluating the health risks from natural toxins, including also the mycotoxins, since 1956. Relevant data for risk assessments used by governments and international risk managers. The Codex Alimentarius Commission is an intergovernmental standards-setting body provides the international reference for national food supplies and for trade in food. In addition, all countries also have own regulatory agencies for risk assessments and legal regulations on mycotoxins in food and feed. Authorities that serve as the scientific advisor to the governmental ministries and other agencies are given in Table 2.

In the context of food safety, and animal and plant health, mycotoxin contaminations have become a global issue. ZEN is a widely distributed mycotoxin, and humans and livestock are exposed to it by the oral intake of contaminated grains, foods, and feeds. The products contaminated with ZEN have potential threatening risks for public health due to its structural similarity to oestrogen. Therefore, it will be focused on ZEN production and ZEN-induced toxicity in the following sections.

## Zearalenone

Zearalenone is a polyketide formerly named as F-2 mycotoxin, and ZEN, ZEA, or ZON are used as abbreviations for it. ZEN, chemically described as *6-[10-hydroxy-6-oxo-trans-1-undecenyl]-β*-resorcyclic acid lactone (Fig. 1a), is an estrogenic metabolite produced by different *Fusarium* species, mainly *F. graminearum* and

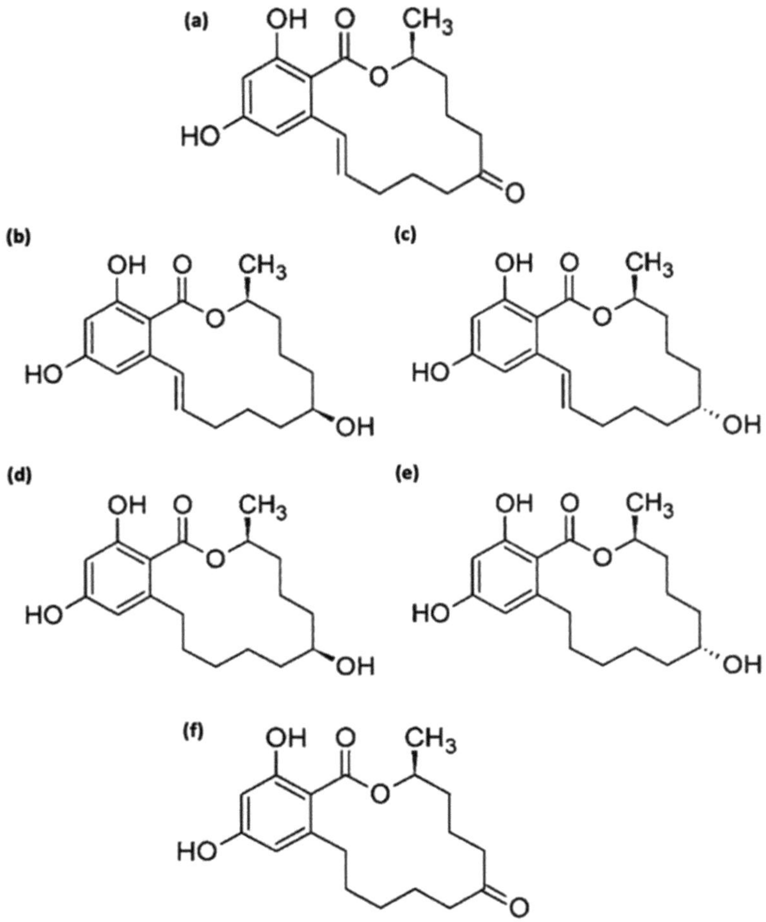

**Fig. 1** Chemical structure of zearalenone ZEN (**a**) and its derivatives: *α*-zearalenol (α-ZEL) (**b**), *β*-zearalenol (β-ZEL) (**c**), *α*-zearalanol (α-ZAL) (**d**) and *β*-zearalanol (β-ZAL) (**e**), zearalanone (ZAN) (**f**) retrieved from (Baggiani et al. 2010)

*F. culmorum* (Atoui et al. 2012). ZEN producers of the genus are *F. equiseti* (Bennett and Klich 2003; Llorens et al. 2022), *F. oxysporum* (Milano and López 1991), *F. verticillioides* (Llorens et al. 2022), *F. cerealis* (synonym, *F. crookwellense*) (Bennett and Klich 2003), *F. semitectum* (Zinedine et al. 2007), and *F. pseudograminearum* (Blaney and Dodman 2002).

Mycotoxins of phytopathogenic *Fusarium* species can be generated due to either field-borne or storage-related infections. Since cereals are the basic source of the human and animal diet, accumulating mycotoxins in food and feed causes serious health problems for humans and animals. ZEN contamination has been reported in different kinds of cereals including corn, wheat, maize, rice, barley, sorghum, and oats (Ropejko and Twarużek 2021). ZEN-producing fungi can contaminate crops in the field, at harvest, and during storage. The production of the toxin may be affected by many factors such as temperature, moisture, substrate, types of isolate and their mutual interactions (Llorens et al. 2004). Economic losses due to mycotoxin contamination are another impact of *Fusarium* toxins in domestic and international feed grain markets (Mielniczuk and Skwaryło-Bednarz 2020). Although the main sources of ZEN are cereals, the presence of ZEN in processed cereal products has been determined. Delicatessen and dairy products are another route for the dietary intake of ZEN and its metabolites (Shikhaliyeva et al. 2020; Ropejko and Twarużek 2021).

## Effects of ZEN Contamination on Organisms

ZEN and its metabolites are classified as non-steroidal estrogen, mycoestrogen, and phytoestrogen, due to their binding affinities to estrogen receptors (Hurd 1977; Shier 1998; Bennett and Klich 2003). They compete with 17-β estradiol to bind the estrogen receptors, and led to disturb the hormonal balance by inhibiting the secretion and release of steroid hormones. Their estrogenic features cause the hyperestrogenic syndrome in laboratory animals, domestic species, and humans (Poór et al. 2015; Zhang et al. 2018). ZEN contamination causes fertility disorders by inducing reproductive tract alterations in animals and humans (Atoui et al. 2012; Janik et al. 2020). In addition, infertility, mummification, and abortions were observed in livestock exposed to ZEN (Abbès et al. 2007; Feng et al. 2008). Among them, pig was reported as the most ZEN-susceptible animal (Chang et al. 2017). In livestock, ZEN can be biotransformed to different kinds of ZEN metabolites. The most occurred derivatives of ZEN are α- and β-zearalenol (α-ZEL and β-ZEL), α- and β-zearalanol (α-ZAL and β-ZAL), and zearalanone (ZAN) (Fig. 1b–f). However, naturally contaminated cereals contain only α-ZAL (Baggiani et al. 2010). α-ZEL, β-ZEL, α-ZAL, and β-ZAL are phase I metabolites. The phase II metabolites are formed due to the conjugation of ZEN and/or its phase I metabolites with glucose or sulfate and glucuronic acid. Those metabolites' amounts and estrogenic activities can vary among different species (EFSA CONTAM Panel (EFSA Panel on Contaminants in the Food Chain) et al. 2017; Ropejko and Twarużek 2021; Llorens et al. 2022; Yli-Mattila et al. 2022). Additionally, it was reported that the estrogenic potencies of ZEN metabolites, except β-ZEL, are higher than ZEN (EFSA CONTAM Panel (EFSA Panel on Contaminants in the Food Chain) et al. 2017).

Cellular responses to ZEN mycotoxin have been studied in experimental animals and cell line systems. ZEN leads to genotoxicity by causing oxidative stress damage as a dose-dependent in experimental animals and different cell lines (Llorens et al. 2022). The hepatotoxic, nephrotoxic, haematotoxic, immunotoxic, and carcinogenic effects of ZEN and its derivatives have also been reported (Cimbalo et al. 2020; Nahle et al. 2021). Several in vivo and in vitro studies provided the identification of various biomarkers that reflect the ZEN toxicity in different tissues and cell types exposed to the mycotoxin (Table 3). Although possible carcinogenic activity

**Table 3** Commonly used biomarkers to identify exposure to ZEN

| Effects of ZEN Exposure | Biomarkers | Studies | References |
|---|---|---|---|
| Reproductive and endocrine effects | Hormonal alterations, relative weight of genital organs | In vivo | Etienne and Jemmali (1982); Takemura et al. (2007); Jiang et al. (2011); Yang et al. (2018) |
| | | In vivo | Yang et al. (2007) |
| Genotoxic effect | Chromosomal abnormalities and inhibition of DNA and protein synthesis | In vivo | Abbès et al. (2007) |
| | | In vivo | Abid-Essefi et al. (2003); Abid-Essefi et al. (2004) |
| Liver damage (the hepatotoxicity of ZEN) | Levels of alanine aminotransferase, aspartate aminotransferase and gamma-glutamyltransferase, and alkaline phosphatase | In vivo | Maaroufi et al. (1996); Jiang et al. (2011) |
| Kidney damage (the nephrotoxicity of ZEN) | The monitoring concentrations of urea nitrogen, uric acid, or creatinine in the blood and urine | In vivo | Jia et al. (2014) |
| Hematologic effect | Analysis of blood parameters such as hematocrit, mean corpuscular volume, number of platelets or white blood cells | In vivo | Etienne and Jemmali (1982); Maaroufi et al. (1996) |
| Cellular response | Evaluation of the superoxide dismutase, glutathione peroxidase, and catalase activity | In vivo | Jiang et al. (2011); Jia et al. (2014) |
| | | In vivo | Jiang et al. (2011); Ben Salem et al. (2015); Tatay et al. (2016) |
| | | In silico | Agahi et al. (2020) |
| | Analysis of malondialdehyde concentrations | In vivo | Jia et al. (2014) |
| | | In vivo | Abid-Essefi et al. (2004); Othmen et al. (2008) |
| | Evaluation of alterations in the expression of antioxidant enzymes and genes related to oxidative stress such as O-6-methylguanine-DNA methyltransferase, glutathione S-transferase, heat shock protein 70, and heme oxygenase-1 | In vivo | Othmen et al. (2008); Karaman et al. (2020) |
| | Assessment of the amount of the cytosolic enzyme lactate dehydrogenase and the lipid profile composition | In vivo | Szabó et al. (2017) |
| | | In vivo | Lei et al. (2013) |

(particularly involvement/association with breast and uterus cancers) of ZEN had been reported, it has been listed as Group 3 carcinogen (not classifiable as to its carcinogenicity to humans) by the International Agency for Research on Cancer (IARC).[1]

## *Fusarium*: ZEN Producer

*Fusarium* genus comprises filamentous ascomycete fungi and is one of the major mycotoxigenic fungi worldwide. The genus contains many agronomically important plant pathogens and opportunistic human pathogens (Ma et al. 2013; King et al. 2015). Its phytopathogens affect global public health by contaminating foodstuffs and feedstuffs with their mycotoxins. *F. graminearum* and *F. culmorum* are the major phytopathogenic species of the genus.

They are predominating causal agents of serious plant diseases (such as Fusarium head blight and crown root rot disease) and lead to economic losses all around the world (Miedaner et al. 2008; Pasquali and Migheli 2014; Matny 2015). *Fusarium* mycotoxins can be grouped as major mycotoxins (including trichothecene, fumonisin, ZEN) and minor mycotoxins (including beauvercin, enniatin, butanolide, equisetin, and fusarin) (Desjardins and Proctor 2007; Yli-Mattila et al. 2022). Different types of mycotoxins accumulate on maize and small grains infected by *Fusarium* spp. In addition to its importance in the agriculture, medicine, and veterinary medicine fields, *Fusarium* species are key model organisms for biological and evolutionary research (Hao et al. 2021). The genome projects for several species have been released on the National Library of Medicine (NIH) genetic sequence database, GenBank. Genome data packages including genome, transcript and protein sequences, annotations, and data reports can be screened in the National Center for Biotechnology Information (NCBI) database. *F. graminearum* PH-1 is the first sequenced reference *Fusarium* genome with a released genome sequence [Genome assembly: ASM24013v3] (King et al. 2015).[2] Also, *F. fujikuroi* IMI 58289, *F. pseudograminearum* CS3096, *F. vanettenii* 77-13-4, *F. verticillioides* 7600, *F. oxysporum* f.sp. *lycopersici* 4287, *F. proliferatum* ET1, *F. odoratissimum* NRRL 54006, *F. venenatum* ASM90000737v1, *F. sporotrichioides* NRRL3299, and *F. mangiferae* MRC7560 genomes were deposited as reference genomes in the NCBI.

In subsequent years, a revised *F. graminearum* genomic reference sequence and genome sequencing and assembly of several other *F. graminearum* strains were released.[3,4] Complete genome assembly and annotation of *F. culmorum* UK99

---

[1] https://monographs.iarc.who.int/list-of-classifications [Retrieved September, 2025]

[2] https://www.ncbi.nlm.nih.gov/data-hub/genome/GCF_000240135.3/ [Retrieved September, 2025]

[3] https://www.ncbi.nlm.nih.gov/data-hub/genome/?taxon=5518 [Retrieved September, 2025]

[4] https://www.ncbi.nlm.nih.gov/genome/browse/#!/eukaryotes/58/ [Retrieved September, 2025]

strain were released. There are three more assemblies for this organism at different assembly levels: contig, scaffold, and chromosome levels.[5] A total of 11.785 genes, 11.444 of which are protein-coding and 341 non-coding, were annotated. The genome size of representative strain Class2-1B was described as 37.5 Mb.[6] Several genes and gene clusters of fungal genomes related to important biological processes available at GenBank provide comprehensive information to perform interspecies comparative analyses and gene function studies. Comparison studies with fungal model organisms including *Neurospora crassa, Magnaporthe grisea, Aspergillus nidulans*, and *F. oxysporum*, in which gene clusters related to mycotoxin biosynthesis are well characterized, provide the biosynthetic pathway characterization in *Fusaria*. Among them, the current status of the *PKS* gene cluster, which is related to the ZEN biosynthetic pathway, is reviewed in this chapter.

## Structure of Core Gene Cluster Related to ZEN Biosynthesis

Polyketides are a major group of fungal secondary metabolites with diverse structures that are involved in various biological activities. They exhibit antibiotic, anticancer, antifungal, antiparasitic, and immunosuppressive properties. Polyketides are synthesized via polyketide synthases (PKSs), one of the multi-domain enzymatic protein families (Staunton and Weissman 2001). Fungal PKSs catalyse the condensation of acetyl-coenzyme A (CoA) and malonyl-CoA during reactions.

Although PKSs have been classified under three groups (types I, -II, and -III), fungal polyketides are formed as type I PKSs. Type I PKSs present in fungal organisms are iterative whereas bacterial ones are modular type I PKSs (Moss et al. 2004). Fungal type I PKSs are multifunctional polypeptides, which have diverse domains to form full-length polyketides. $\beta$-ketoacyl synthase (KS), acyl transferase (AT), and acyl carrier protein (ACP) [also called as a phosphopantetheine attachment site (PP) domains] are essential structures for PKSs. Some of PKSs have additional catalytic domains which participate in $\beta$-carbonyl processing such as $\beta$-ketoacyl reductase (KR), dehydratase (DH) and enoyl reductase (ER) (Kroken et al. 2003).

Genes that have functions in the biosynthetic pathway of a secondary metabolite often are located as a cluster in fungi genome. One or more regulatory genes found in the cluster participate in and regulate biosynthesis (Gaffoor and Trail 2006). 15 PKSs have been encoded from *F. graminearum* genome (Gaffoor et al. 2005). Among them, only PKS4 and PKS13 have a function for ZEN biosynthesis. ZEN has been synthesized by the combinational action of these two PKSs, an isoamyl alcohol oxidase and a transcriptional regulator, encoding from the *PKS* gene cluster.

[5] https://www.ncbi.nlm.nih.gov/data-hub/genome/?taxon=5516 [Retrieved September, 2025]

[6] https://www.ncbi.nlm.nih.gov/data-hub/genome/GCA_016952355.1/ [Retrieved September, 2025]

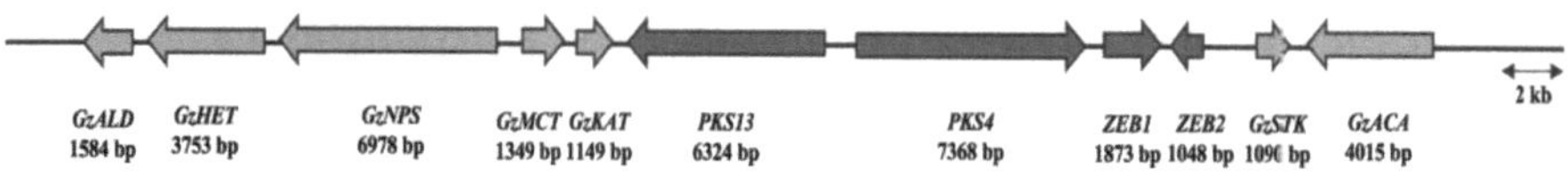


**Fig. 2** The genomic structure of the *PKS* gene cluster with 49.993 kb DNA region. Genes located in the cluster and the direction of their transcription showed by the arrows and arrowheads, respectively. Dark-filled arrows indicate the identified genes essential in the ZEN biosynthetic pathway

**Table 4** Genes related to ZEN production

| Gene | Protein ID | Protein | Length (bp) | Exon Number | Strand |
|---|---|---|---|---|---|
| *PKS13* | ABB90282.1 | Polyketide synthase | 6324 | 5 | Minus |
| *PKS4* | ABB90283.1 | Polyketide synthase | 7368 | 6 | Plus |
| *ZEB1* | ABB90284.1 | Isoamyl alcohol oxidase | 1873 | 4 | Plus |
| *ZEB2* | ABB90285.1 | Transcription factor with bZIP domain | 1048 | 2 | Minus |

The biosynthetic gene cluster consists of totally 11 genes (please see NCBI GenBank accession number: DQ019316.1), of which two are iterative fungal PKS genes (*PKS4* and *PKS13*) together with *ZEB1* (ZEN biosynthesis gene 1) and *ZEB2* (– gene 2), which are required for ZEN production (Fig. 2).

*ZEB1* encodes an isoamyl alcohol oxidase. *ZEB2* codes for a transcription factor. ZEN biosynthetic genes occupy a cluster with the 49.993 bp length at the cluster (Kim et al. 2005), and are located across four contigs (the contig numbers: 1.117,1.118, 1.119 and 1.120) on chromosome 1 of *F. graminearum* PH-1. *PKS4*, *PKS13*, *ZEB1* and *ZEB2* genes are involved in a four-gene-core cluster of 18.433 bp (Table 4). PKS4 protein includes KS, AT, DH, ER, KR, and ACP (also known as PP domains). PKS13 possesses conserved structures for PKSs including KS, AT, ACP, PP domains, and other domains: starter unit acyltransferase, product template, and thioesterase domains (Kroken et al. 2003; Hansen et al. 2015). ZEB1 and ZEB2 have common domains (isoamyl alcohol oxidase and DNA binding motif, respectively) well-defined on GenBank and Swissmodel/expasy databases.

Open reading frames (ORF) of *ZEB1* [FG12056], *ZEB2* [FG02398], *GzSTK* [FG02399], and *GzACA* [FG02400] genes were located at the 3′ region of the *PKS4*[7] [FG12126]. *GzKAT* [FG12015], *GzMCT* [FG13438], *GzNPS* [FG02394], *GzHET* [FG02393], and *GzALD* [FG02392] were located downstream of the *PKS13*[8] [FG02395] (Mewes et al. 2004; Kim et al. 2005; Güldener et al. 2006).

[7] *PKS4* gene were also called as ZEA2 (Gaffoor and Trail 2006).

[8] *PKS13* gene were also called as ZEA1 (Gaffoor and Trail 2006).

## PKS4 *and* PKS13*: Essential Genes in ZEN Biosynthesis*

Techniques used in advanced molecular analysis such as targeted gene disruption, overexpression, etc. have provided important data to understand the mechanism of ZEN biosynthesis. PKS4 initiates the ZEN biosynthetic pathway. It proceeds the reduction reactions, resulting in hexaketide and this moiety passed onto the non-reducing PKS13, which triggers the gene expression for further elongation of the polyketide chain (Fig. 3) (Gaffoor and Trail 2006). It is predicted that the ER domain is required for ZEN biosynthesis. Although eight of the PKS proteins have ER domain, it was reported that only *PKS4* was upregulated during the ZEN contamination *in planta* (Malz et al. 2005; Lysøe et al. 2006). Lysøe et al. (2006) reported that ZEN was not produced when the central *PKS4* was replaced with hygromycin phosphotransferase (*hygB*) marker gene in *F. graminearum*.

Moreover, no pathogenicity for *PKS4* deleted mutants was detected via barley infection studies. The comparison of PKS sequences between ZEN producer and non-ZEN producers showed that PKS4 was one of the proteins which contain ER domains in *F. graminearum*. Moreover, it was reported that PKS4 has not any known orthologue. The correlation of gene manipulation findings with transcriptional and bioinformatic analysis data can be used as proof of the role of *PKS4* in ZEN production. ZEN and its derivatives do not affect the transcriptional level of the *PKS4* gene. The fact that targeted *PKS4* gene replacement resulted in the down-regulation of the *PKS13* gene indicates the stimulation effect of PKS4 on the gene expression of *PKS13* (Kim et al. 2005; Lysøe et al. 2006).

Deletion of both *PKS4* and *PKS13* caused the reduction of transcript levels of *ZEB1* and *ZEB2* together with additional genes -*GzSTK, GzKAT, GzMCT,* and *GzNPS*- which are located in the *PKS* gene cluster. Moreover, it has been determined that the disruption of *PKS13* ORF led to the reduction of *PKS4* transcription and also the non-production of ZEN (Kim et al. 2005).

It was revealed that the expression level of *PKS13* was higher than *PKS4*. Independently transcriptions of both genes or transcripts occurring from the shared promotor resulting in various transcript sizes (longer transcript size of *PKS4*) could be the reasons for the production of different expression profiles (Gaffoor and Trail 2006; Lysøe et al. 2008).

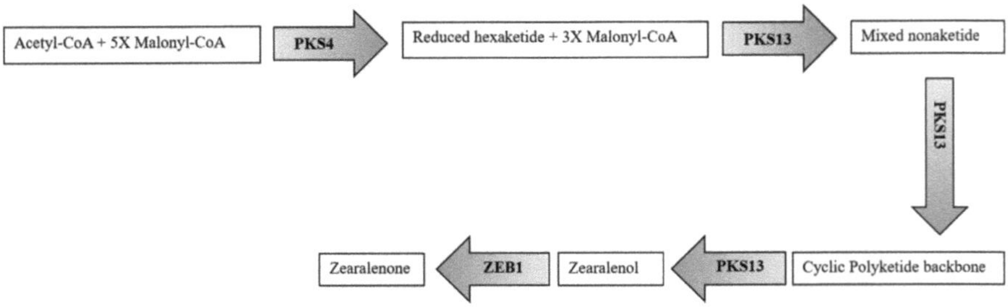


**Fig. 3** ZEN biosynthetic pathway

*ZEB1* and *ZEB2* encode proteins, which are similar to isoamyl alcohol oxidase, and a basic leucine zipper (bZIP) transcription factor (TF), respectively (Kim et al. 2005; Gaffoor and Trail 2006; Lysøe et al. 2006). ZEB1 protein catalyses an oxidation step for converting *β*-zearalenol (*β*-ZOL) to ZEA in *G. zeae*. *ZEB1* deletions did not dramatically affect the transcription of *PKS4*, *PKS13*, and additional members of cluster: *GzSTK*, *GzKAT*, *GzMCT*. However, *ZEB2* and *GzNPS* were strongly reduced by *ZEB1* deletion (Kim et al. 2005).

ZEB2 controls the activity of three genes *PKS4*, *PKS13*, and *ZEB1* (Lysøe et al. 2008). ZEB2L and ZEB2S are the alternative transcripts of *ZEB2*, using alternative promotors. Both transcripts are translated into functional proteins. Because ZEB2L contains a bZIP DNA binding domain, it influences the promoter region of the ZEN biosynthetic cluster. However, the bZIP domain is not carried by ZEB2S. When both transcripts are co-expressed, ZEB2S reduces the affinity of ZEB2L to the promoter. No more ZEN production depending on the co-expression of both transcripts and the accumulation of ZEN at a high level depending on only ZEB2L expression reveal that ZEB2S is a regulator protein on inactivation of transcriptional activator ZEB2L. In conclusion, *ZEB2* isoforms (ZEB2S and ZEB2L), caused by alternative promoter usage and feedback loop mechanism, are responsible for the autoregulation of the ZEN biosynthetic pathway (Park et al. 2015; Kim et al. 2018).

## Concluding Remarks

Uncovering the *PKS* gene cluster and ZEN biosynthetic pathway contributes to improving strategies implemented for the screening and diagnosing ZEN-producing species. Knowledge of the *PKS* genes and their regulation will provide important data to understand the basis of fungal secondary metabolites and contribute to reducing the impact of mycotoxin contamination on society by designing projects for the prevention, control, and mitigation of fungal contamination. The data about the four-core gene required for the ZEN biosynthesis have been accessed under the accession number DQ019316.1 on GenBank.[9] *PKS* gene cluster corresponds to 49.993 bp region. It includes additional seven genes (*GzALD*, *GzHET*, *GzNPS*, *GzMCT*, *GzKAT*, *GzSTK*, and *GzACA*), which have been indirectly associated with ZEN biosynthesis (Table 5) (Kim et al. 2005). There is limited knowledge about the linkage between identified ZEN biosynthetic genes and other seven genes in the same cluster. Further omic studies should be carried out to generate integrative data about the regulation of the *PKS* gene cluster. Outcomes of further biochemical and bioinformatic analysis targeting these seven additional genes are vital for determining the precise functions of these genes in ZEN biosynthesis. Additionally, investigations of fungal *PKS* genes could be helpful in producing new compounds using engineering methods.

[9] https://www.ncbi.nlm.nih.gov/nuccore/DQ019316.1 [Retrieved September, 2025]

**Table 5** Additional genes located at *PKS* gene cluster

| Gene | Protein ID | Protein | Length (bp) | Exon Number | Strand |
|---|---|---|---|---|---|
| *GzALD* | ABB90277.1 | Aldehyde dehydrogenase | 1584 | 3 | Minus |
| *GzHET* | ABB90278.1 | Heterokaryon incompatibility protein | 3753 | 1 | Minus |
| *GzNPS* | ABB90279.1 | Non-ribosomal peptide synthetase | 6978 | 1 | Minus |
| *GzMCT* | ABB90280.1 | Monocarboxylate transporter like protein | 1349 | 4 | Plus |
| *GzKAT* | ABB90281.1 | K+ channel protein | 1149 | 3 | Plus |
| *GzSTK* | ABB90286.1 | Protein kinase Eg2-like protein | 1090 | 2 | Plus |
| *GzACA* | ABB90287.1 | $Ca^{2+}$ ATPase | 4015 | 3 | Minus |

# References


Abbès S, Ouanes Z, Salah-Abbès JB, Abdel-Wahhab MA, Oueslati R, Bacha H (2007) Preventive role of aluminosilicate clay against induction of micronuclei and chromosome aberrations in bone-marrow cells of Balb/c mice treated with Zearalenone. Mutat Res/Genet Toxicol Environ Mutag 631(2):85–92. https://doi.org/10.1016/j.mrgentox.2007.01.012

Abid-Essefi S, Baudrimont I, Hassen W, Ouanes Z, Mobio TA, Anane R, Creppy EE, Bacha H (2003) DNA fragmentation, apoptosis and cell cycle arrest induced by zearalenone in cultured DOK, Vero and Caco-2 cells: prevention by vitamin E. Toxicology 192(2–3):237–248. https://doi.org/10.1016/S0300-483X(03)00329-9

Abid-Essefi S, Ouanes Z, Hassen W, Baudrimont I, Creppy E, Bacha H (2004) Cytotoxicity, inhibition of DNA and protein syntheses and oxidative damage in cultured cells exposed to zearalenone. Toxicol In Vitro 18(4):467–474. https://doi.org/10.1016/j.tiv.2003.12.011

Agahi F, Juan C, Font G, Juan-García A (2020) *In silico* methods for metabolomic and toxicity prediction of zearalenone, α-zearalenone and β-zearalenone. Food Chem Toxicol 146:111818. https://doi.org/10.1016/j.fct.2020.111818

Atoui A, El Khoury A, Kallassy M, Lebrihi A (2012) Quantification of *Fusarium graminearum* and *Fusarium culmorum* by real-time PCR system and zearalenone assessment in maize. Int J Food Microbiol 154(1–2):59–65. https://doi.org/10.1016/j.ijfoodmicro.2011.12.022

Baggiani C, Anfossi L, Giovannoli C (2010) Artificial systems for molecular recognition of mycotoxins. In: Mycotoxins in food, feed and bioweapons. Springer, Berlin, Heidelberg, pp 3–20

Ben Salem I, Boussabbeh M, Helali S, Abid-Essefi S, Bacha H (2015) Protective effect of Crocin against zearalenone-induced oxidative stress in liver and kidney of Balb/c mice. Environ Sci Pollut Res 22(23):19069–19076. https://doi.org/10.1007/s11356-015-5086-2

Bennett J, Klich M (2003) Mycotoxins. Clin Microbiol Rev 16:497–516. https://doi.org/10.1128/CMR.16.3.497-516.2003

Blaney BJ, Dodman RL (2002) Production of zearalenone, deoxynivalenol, nivalenol, and acetylated derivatives by Australian isolates of *Fusarium graminea*rum and *F. pseudograminearum* in relation to source and culturing conditions. Aust J Agric Res 53:1317–1326. https://doi.org/10.1071/AR02041

Chang H, Kim W, Park JH, Kim D, Kim CR, Chung S, Lee C (2017) The occurrence of zearalenone in South Korean feedstuffs between 2009 and 2016. Toxins 9(7):223. https://doi.org/10.3390/toxins9070223

Cimbalo A, Alonso-Garrido M, Font G, Manyes L (2020) Toxicity of mycotoxins *in vivo* on vertebrate organisms: a review. Food Chem Toxicol 137:111161. https://doi.org/10.1016/j.fct.2020.111161

Desjardins AE, Proctor RH (2007) Molecular biology of *Fusarium* mycotoxins. Int J Food Microbiol 119(1–2):47–50. https://doi.org/10.1016/j.ijfoodmicro.2007.07.024

EFSA CONTAM Panel (EFSA Panel on Contaminants in the Food Chain), Knutsen HK, Alexander J, Barregård L, Bignami M, Brüschweiler B, Ceccatelli S, Cottrill B, Dinovi M, Edler L, Grasl-Kraupp B, Hogstrand C, Hoogenboom LR, Nebbia CS, Petersen A, Rose M, Roudot AC, Schwerdtle T, Vleminckx C, Vollmer G, Wallace H, Dall'Asta C, Dänicke S, Eriksen GS, Altieri A, Roldán-Torres R, Oswald IP (2017) Risks for animal health related to the presence of zearalenone and its modified forms in feed. EFSA J 15(7):4851, 123. https://doi.org/10.2903/j.efsa.2017.4851

Etienne M, Jemmali M (1982) Effects of zearalenone (F2) on estrous activity and reproduction in gilts. J Anim Sci 55(1):1–10. https://doi.org/10.2527/jas1982.5511

Feng J, Shan M, Du H, Han X, Xu Z (2008) *In vitro* adsorption of zearalenone by cetyltrimethyl ammonium bromide-modified montmorillonite nanocomposites. Microporous Mesoporous Mater 113(1–3):99–105. https://doi.org/10.1016/j.micromeso.2007.11.007

Gaffoor I, Trail F (2006) Characterization of two polyketide synthase genes involved in zearalenone biosynthesis in *Gibberella zeae*. Appl Environ Microbiol 72(3):1793–1799. https://doi.org/10.1128/AEM.72.3.1793-1799.2006

Gaffoor I, Brown DW, Plattner R, Proctor RH, Qi W, Trail F (2005) Functional analysis of the polyketide synthase genes in the filamentous fungus *Gibberella zeae* (anamorph *Fusarium graminearum*). Eukaryot Cell 4(11):1926–1933. https://doi.org/10.1128/EC.4.11.1926-1933.2005

Güldener U, Mannhaupt G, Münsterkötter M, Haase D, Oesterheld M, Stümpflen V, Meves H-W, Adam G (2006) FGDB: a comprehensive fungal genome resource on the plant pathogen *Fusarium graminearum*. Nucleic Acids Res 34:456–458. https://doi.org/10.1093/nar/gkj026

Gurikar C, Shivaprasad DP, Sabillón L, Gowda NN, Siliveru K (2023) Impact of mycotoxins and their metabolites associated with food grains. Grain Oil Sci Technol 6(1):1–9. https://doi.org/10.1016/j.gaost.2022.10.001

Hansen FT, Gardiner DM, Lysøe E, Fuertes PR, Tudzynski B, Wiemann P, Sondergaard TE, Giesedi H, Brodersen DE, Sørensen JL (2015) An update to polyketide synthase and non-ribosomal synthetase genes and nomenclature in *Fusarium*. Fungal Genet Biol 75:20–29. https://doi.org/10.1016/j.fgb.2014.12.004

Hao Z, Li Y, Jiang Y, Xu J, Li J, Luo L (2021) Genome sequence analysis of the fungal pathogen *Fusarium graminearum* using oxford nanopore technology. J Fungi 7(9):699. https://doi.org/10.3390/jof7090699

Hurd RN (1977) Structure activity relationships in zearalenones. In: Rodricks JV, Hesseltine CW, Mehlman MA (eds) Mycotoxins in human and animal health. Pathotox Publications, Inc., Park Forest South, pp 379–391

Janik E, Niemcewicz M, Ceremuga M, Stela M, Saluk-Bijak J, Siadkowski A, Bijak M (2020) Molecular aspects of mycotoxins—a serious problem for human health. Int J Mol Sci 21(21):8187. https://doi.org/10.3390/ijms21218187

Jia Z, Liu M, Qu Z, Zhang Y, Yin S, Shan A (2014) Toxic effects of zearalenone on oxidative stress, inflammatory cytokines, biochemical and pathological changes induced by this toxin in the kidney of pregnant rats. Environ Toxicol Pharmacol 37(2):580–591. https://doi.org/10.1016/j.etap.2014.01.010

Jiang SZ, Yang ZB, Yang WR, Gao J, Liu FX, Broomhead J, Chi F (2011) Effects of purified zearalenone on growth performance, organ size, serum metabolites, and oxidative stress in postweaning gilts. J Anim Sci 89(10):3008–3015. https://doi.org/10.2527/jas.2010-3658

Karaman EF, Ariman I, Ozden S (2020) Responses of oxidative stress and inflammatory cytokines after zearalenone exposure in human kidney cells. World Mycotoxin J 13(3):411–421. https://doi.org/10.3920/WMJ2019.2512

Kim YT, Lee YR, Jin J, Han KH, Kim H, Kim JC, Lee T, Yun SH, Lee YW (2005) Two different polyketide synthase genes are required for synthesis of zearalenone in *Gibberella zeae*. Mol Microbiol 58(4):1102–1113. https://doi.org/10.1111/j.1365-2958.2005.04884.x

Kim JE, Son H, Lee YW (2018) Biosynthetic mechanism and regulation of zearalenone in *Fusarium graminearum*. JSM Mycotoxins 68(1):1–6. https://doi.org/10.2520/myco.68-1-2

King R, Urban M, Hammond-Kosack MC, Hassani-Pak K, Hammond-Kosack KE (2015) The completed genome sequence of the pathogenic ascomycete fungus *Fusarium graminearum*. BMC Genomics 16(1):544. https://doi.org/10.1186/s12864-015-1756-1

Kroken S, Glass NL, Taylor JW, Yoder OC, Turgeon BG (2003) Phylogenomic analysis of type I polyketide synthase genes in pathogenic and saprobic ascomycetes. PNAS 100(26):15670–15675. https://doi.org/10.1073/pnas.2532165100

Kuhn DM, Ghannoum M (2003) Indoor mold, toxigenic fungi, and *Stachybotrys chartarum*: infectious disease perspective. Clin Microbiol Rev 16(1):144–172. https://doi.org/10.1128/CMR.16.1.144-172.2003

Lei M, Zhang N, Qi D (2013) *In vitro* investigation of individual and combined cytotoxic effects of aflatoxin B1 and other selected mycotoxins on the cell line porcine kidney 15. Exp Toxicol Pathol 65(7–8):1149–1157. https://doi.org/10.1016/j.etp.2013.05.007

Llorens A, Mateo R, Hinojo MJ, Logrieco A, Jimenez M (2004) Influence of the interactions among ecological variables in the characterization of zearalenone producing isolates of *Fusarium* spp. Syst Appl Microbiol 27(2):253–260. https://doi.org/10.1078/072320204322881871

Llorens P, Herrera M, Juan-García A, Payá JJ, Moltó JC, Ariño A, Juan C (2022) Biomarkers of exposure to zearalenone in *in vivo* and *in vitro* studies. Toxins 14(5):291. https://doi.org/10.3390/toxins14050291

Lysøe E, Klemsdal SS, Bone KR, Frandsen RJ, Johansen T, Thrane U, Giese H (2006) The *PKS4* gene of *Fusarium graminearum* is essential for zearalenone production. Appl Environ Microbiol 72(6):3924–3932. https://doi.org/10.1128/AEM.00963-05

Lysøe E, Bone KR, Klemsdal SS (2008) Identification of up-regulated genes during zearalenone biosynthesis in *Fusarium*. Eur J Plant Pathol 122(4):505–516. https://doi.org/10.1007/s10658-008-9318-x

Ma LJ, Geiser DM, Proctor RH, Rooney AP, O'Donnell K, Trail F, Gardiner DM, Manners JM, Kazan K (2013) *Fusarium* pathogenomics. Ann Rev Microbiol 67:399–416. https://doi.org/10.1146/annurev-micro-092412-155650

Maaroufi K, Chekir L, Creppy EE, Ellouz F, Bacha H (1996) Zearalenone induces modifications of haematological and biochemical parameters in rats. Toxicon 34(5):535–540. https://doi.org/10.1016/0041-0101(96)00008-6

Mahato DK, Devi S, Pandhi S, Sharma B, Maurya KK, Mishra S, Dhawan K, Selvakumar R, Kamle M, Mishra AK, Kumar P (2021) Occurrence, impact on agriculture, human health, and management strategies of zearalenone in food and feed: a review. Toxins 13(2):92. https://doi.org/10.3390/toxins13020092

Malz S, Grell MN, Thrane C, Maier FJ, Rosager P, Felk A, Albertsen KS, Salomon S, Bohm L, Schäfer W, Giese H (2005) Identification of a gene cluster responsible for the biosynthesis of aurofusarin in the *Fusarium graminearum* species complex. Fungal Genet Biol 42(5):420–433. https://doi.org/10.1016/j.fgb.2005.01.010

Murray PR, Rosenthal KS, Pfaller MA (2020) Medical microbiology E-book. Elsevier Health Sciences

Smith M, McGinnis MR (2009) Mycotoxins and their effects on humans. In: Clinical mycology. Churchill Livingstone, pp 649–656

Peraica M (2016) Mycotoxicoses. In: Environmental mycology in public health. Academic Press, pp 45–49

Rogowska A, Pomastowski P, Sagandykova G, Buszewski B (2019) Zearalenone and its metabolites: effect on human health, metabolism and neutralisation methods. Toxicon 162:46–56. https://doi.org/10.1016/j.toxicon.2019.03.004

Zain ME (2011) Impact of mycotoxins on humans and animals. J Saudi Chem Soc 15(2):129–144. https://doi.org/10.1016/j.jscs.2010.06.006

Milano GD, López TA (1991) Influence of temperature on zearalenone production by regional strains of *Fusarium graminearum* and *Fusarium oxysporum* in culture. Int J Food Microbiol 13(4):329–333. https://doi.org/10.1016/0168-1605(91)90092-4

Zinedine A, Soriano JM, Molto JC, Manes J (2007) Review on the toxicity, occurrence, metabolism, detoxification, regulations and intake of zearalenone: an oestrogenic mycotoxin. Food Chem Toxicol 45(1):1–18. https://doi.org/10.1016/j.fct.2006.07.030

Ropejko K, Twarużek M (2021) Zearalenone and its metabolites—general overview, occurrence, and toxicity. Toxins 13(1):35. https://doi.org/10.3390/toxins13010035

Mielniczuk E, Skwaryło-Bednarz B (2020) Fusarium head blight, mycotoxins and strategies for their reduction. Agronomy 10(4):509. https://doi.org/10.3390/agronomy10040509

Shikhaliyeva I, Teker T, Albayrak G (2020) Masked mycotoxins of deoxynivalenol and zearalenone—unpredicted toxicity. Biomed J Sci Tech Res 29(2):22288–22293. https://doi.org/10.26717/BJSTR.2020.29.004773

Shier WT (1998) Estrogenic mycotoxins. Revue de Médecine Vétérinaire 149(6):599–604

Poór M, Kunsági-Máté S, Sali N, Kőszegi T, Szente L, Peles-Lemli B (2015) Interactions of zearalenone with native and chemically modified cyclodextrins and their potential utilization. J Photochem Photobiol B Biol 151:63–68. https://doi.org/10.1016/j.jphotobiol.2015.07.009

Zhang GL, Feng YL, Song JL, Zhou XS (2018) Zearalenone: a mycotoxin with different toxic effect in domestic and laboratory animals' granulosa cells. Front Genet 9:667. https://doi.org/10.3389/fgene.2018.00667

Yli-Mattila T, Yörük E, Abbas A, Teker T (2022) Overview on major mycotoxins accumulated on food and feed. In: Fungal biotechnology: prospects and avenues, vol 310. https://doi.org/10.1201/9781003248316-16

Nahle S, El Khoury A, Atoui A (2021) Current status on the molecular biology of zearalenone: its biosynthesis and molecular detection of zearalenone producing *Fusarium* species. Eur J Plant Pathol 159(2):247–258. https://doi.org/10.1007/s10658-020-02173-9

Takemura H, Shim JY, Sayama K, Tsubura A, Zhu BT, Shimoi K (2007) Characterization of the estrogenic activities of zearalenone and zeranol *in vivo* and *in vitro*. J Steroid Biochem Mol Biol 103(2):170–177. https://doi.org/10.1016/j.jsbmb.2006.08.008

Yang LJ, Zhou M, Huang LB, Yang WR, Yang ZB, Jiang SZ, Ge JS (2018) Zearalenone-promoted follicle growth through modulation of Wnt-1/β-catenin signaling pathway and expression of estrogen receptor genes in ovaries of postweaning piglets. J Agric Food Chem 66(30):7899–7906. https://doi.org/10.1021/acs.jafc.8b02101

Yang J, Zhang Y, Wang Y, Cui S (2007) Toxic effects of zearalenone and α-zearalenol on the regulation of steroidogenesis and testosterone production in mouse Leydig cells. Toxicol In Vitro 21(4):558–565. https://doi.org/10.1016/j.tiv.2006.10.013

Tatay E, Font G, Ruiz MJ (2016) Cytotoxic effects of zearalenone and its metabolites and antioxidant cell defense in CHO-K1 cells. Food Chem Toxicol 96:43–49. https://doi.org/10.1016/j.fct.2016.07.027

Othmen ZOB, El Golli E, Abid-Essefi S, Bacha H (2008) Cytotoxicity effects induced by Zearalenone metabolites, α Zearalenol and β Zearalenol, on cultured Vero cells. Toxicology 252(1–3):72–77. https://doi.org/10.1016/j.tox.2008.07.065

Szabó A, Szabó-Fodor J, Fébel H, Mézes M, Balogh K, Bázár G, Kocsó D, Ali O, Kovács M (2017) Individual and combined effects of fumonisin B1, deoxynivalenol and zearalenone on the hepatic and renal membrane lipid integrity of rats. Toxins 10(1):4. https://doi.org/10.3390/toxins10010004

Miedaner T, Cumagun CJR, Chakraborty S (2008) Population genetics of three important head blight pathogens *Fusarium graminearum*, *F. pseudograminearum* and *F. culmorum*. J Phytopathol 156(3):129–139. https://doi.org/10.1111/j.1439-0434.2007.01394.x

Pasquali M, Migheli Q (2014) Genetic approaches to chemotype determination in type B-trichothecene producing *Fusaria*. Int J Food Microbiol 189:164–182. https://doi.org/10.1016/j.ijfoodmicro.2014.08.011

Matny ON (2015) Fusarium head blight and crown rot on wheat & barley: losses and health risks. Adv Plants Agric Res 2(1):2–7. https://doi.org/10.15406/apar.2015.02.00039

Staunton J, Weissman KJ (2001) Polyketide biosynthesis: a millennium review. Nat Prod Rep 18(4):380–416. https://doi.org/10.1039/A909079G

Moss SJ, Martin CJ, Wilkinson B (2004) Loss of co-linearity by modular polyketide synthases: a mechanism for the evolution of chemical diversity. Nat Prod Rep 21(5):575–593. https://doi.org/10.1039/B315020H

Mewes HW, Amid C, Arnold R, Frishman D, Güldener U, Mannhaupt G, Münsterkötter M, Pagel P, Strack N, Stümpflen V, Warfsmann J, Ruepp A (2004) MIPS: analysis and annotation of proteins from whole genomes. Nucleic Acids Res 32:D41–D44. https://doi.org/10.1093/nar/gkh092

Park AR, Son H, Min K, Park J, Goo JH, Rhee S, Chae SK, Lee YW (2015) Autoregulation of *ZEB2* expression for zearalenone production in *Fusarium graminearum*. Mol Microbiol 97(5):942–956. https://doi.org/10.1111/mmi.13078

# Regulation of *Tri5* Gene Cluster in *Fusarium* Species Through *tri4* and *tri5* Genes

**Gülruh Albayrak, Gülin İnci Varol, and Tuğba Teker**

## Introduction

Mycotoxins are the secondary metabolites with low molecular weight produced by fungi. They can have adverse effects on plant, human, and animal health. Whereas it is claimed that many fungal genera contain mycotoxigenic species and hundreds of different mycotoxin species have already been identified, predominant mycotoxins are produced by mainly *Aspergillus, Penicillium*, and *Fusarium* genera (Frisvad et al. 2011; Ismaiel and Papenbrock 2015).

Some of these mycotoxins are synthesized by several species belonging to the same genus, while others are by the members of phylogenetically different genera (Frisvad et al. 2011). Among them, the genus *Fusarium* consists of at least 300 distinct species, approximately 70 of which were well-known (O'Donnell et al. 2018; Summerell 2019). Toxigenic *Fusarium* species can be infectious on a wide variety of crops, including all small grains, banana, tobacco, and carnation, and are the causal agents of certain diseases such as Fusarium Head Blight, crown root rot and Fusarium wilt (Moretti 2009; LaMondia 2015; Basallote-Ureba et al. 2016). Contamination with *Fusarium* spp. and further accumulation of mycotoxins resulted in severe yield losses, and crop quality reduction arose from necrosis, chlorosis, and mortality on a cellular basis (Sudakin 2003; Saharan et al. 2004; Miedaner et al. 2008).

G. Albayrak (✉) · G. İ. Varol
Science Faculty, Department of Molecular Biology and Genetics, Istanbul University, Vezneciler, Istanbul, Türkiye
e-mail: gulruh@istanbul.edu.tr; gulin.varol@istanbul.edu.tr

T. Teker
Faculty of Engineering and Natural Sciences, Department of Molecular Biology and Genetics, Istanbul Atlas University, Kağıthane, Istanbul, Türkiye
e-mail: tugba.teker@atlas.edu.tr

 
A. Gomes Rodrigues, A. Gupta (eds.), *Fungal Infections*, Fungal Biology,
https://doi.org/10.1007/978-3-032-06014-3_5

The consumption of accumulated mycotoxins in plants causes some significant health problems named mycotoxicosis for humans and animals (Sáenz et al. 2020). In addition to the main types of mycotoxins, their modified derivatives, named masked mycotoxins, were consumed by humans and animals. Masked forms, detoxification products of plants, maintain the harmful effects on humans and animals since they were converted to their main forms by the intestinal microbiome. Therefore, mycotoxins and their producers must be regularly considered to generate risk assessment protocols (Berthiller et al. 2013).

Additionally, some *Fusarium* species have been reported as fungicolous fungi, which refers to a diverse group associated with or even become infectious to other fungi (Sun et al. 2019; Torbati et al. 2021). It has been described that several hundred compounds have been toxic or potentially toxic secondary metabolites produced by *Fusarium* spp. Trichothecenes are known as the most destructive mycotoxins due to their easy and high accumulation potentials in organisms when consumed through the food chain. Also, because of their ability to interact with the eukaryotic ribosomes and to prevent polypeptide chain synthesis during initiation or elongation steps of translation, studies about mycotoxins and pathogenic fungi become remarkable (Cundliffe et al. 1974).

It is important to understand the structure, pathogenesis, mode of action, and regulation strategies for the microorganism and its toxic secondary metabolites to provide effective control strategies. First of all, knowledge about trichothecene structure, classification, and biosynthesis was given in this chapter. Then pathogenesis of *Fusarium* spp. has been compiled by examining trichothecenes' cellular effects. Importantly, the results of mycotoxin accumulation in organism and *Fusarium* consumption are scrutinized from a clinical point of view. Furthermore, their impacts on targets were highlighted by clarifying both cellular mechanisms and organism-scale sickness patterns. Finally, regulation studies on the trichothecene production-related genes and their importance were given.

## Trichothecenes and Their Classification

Trichothecenes are mycotoxins with a tetracyclic sesquiterpenoid structure that has more than 200 different forms with various toxicities (Chen et al. 2019a). In addition to the *Fusarium* genus, they are produced by *Myrothecium, Spicellum, Stachybotrys, Cephalosporium, Trichoderma,* and *Trichothecium* genera (Proctor et al. 2020). All trichothecenes share a common chemical skeleton. Since the molecule is small and amphipathic, its structure gives moving ability across the cell membrane passively and easily. In addition, the absorption and ingestion of trichothecenes become rapid in cells due to their skeleton (McCormick et al. 2011). The skeleton, which constitutes three main ring structures, is 12,13-epoxytrichothec-9-ene core. One of these ring structures, tetrahydropyran ring, is responsible for the toxigenic property of trichothecenes. Conformational differences and substitutional properties, derived from the tetracyclic ring system, resulted in different types of trichothecenes (Ismaiel and Papenbrock 2015).

Trichothecenes produced by *Fusarium* have distinguishable characteristics from the trichothecenes produced by other species. They have an additional hydroxyl or acetyl group at C-3. Meanwhile, the other species can produce trichothecene without this functional group (McCormick et al. 2011). Trichothecenes are classified into four groups; type A, -B, -C, and -D based on the substitution pattern of the skeleton (Ismaiel and Papenbrock 2015).

Type A trichothecenes have a hydroxyl or ester group only at C-8. They are positioned as the simplest type of trichothecene. Neosolaniol, T-2 toxin, trichodermin, NX-2, diacetoxyscirpenol (DAS), and harzianum are included in this group (Foroud and Eudes 2009; Varga et al. 2015). A ketone group positioned at C-8 and a hydroxyl group at C-7 present only in *Fusarium* spp., differentiates type B trichothecenes from others. Deoxynivalenol (DON), nivalenol (NIV), and their acetylated derivatives (3-ADON, 15-ADON, 4-ANIV) are classified in this group.

The last two classes are Type C and -D trichothecenes which cannot be produced by *Fusarium* spp. Crotocin becomes a type C trichothecene. This class is distinguished via epoxide ring at C-7 or C-8. The last group of trichothecenes is type D and its distinctive feature is having a cyclic diester or triester linkage between C-4 and C-15. Roridin A, verrucarin A, and satratoxin H are included in this class (McCormick et al. 2011; Ismaiel and Papenbrock 2015; Chen et al. 2019a). The detailed substitution pattern of type A and type B trichothecenes, which are able to be produced by *Fusarium* spp., is given in Table 1.

**Table 1** Functional groups in core structure of type A and type B trichothecenes produced by *Fusarium* spp. –OAc: acetyl; –OIsoval: isovalerate. (From Foroud and Eudes 2009)

| | C3 | C4 | C7 | C8 | C15 |
|---|---|---|---|---|---|
| *Type A* | | | | | |
| Diacetoxyscirpenol (DAS) | –OH | –OAc | –H | –H | –OAc |
| Trichodermin | –H | –OAc | –H | –H | –H |
| Trichodermol | –H | –OH | –H | –H | –H |
| T-2 toxin | –OH | –OAc | –H | –OIsoval | –OAc |
| HT-2 toxin | –OH | –OH | –H | –OIsoval | –OAc |
| NX-2 | –OAc | –H | –OH | –H | –OH |
| NX-3 | –OH | –H | –OH | –H | –OH |
| *Type B* | | | | | |
| Nivalenol (NIV) | –OH | –OH | –OH | =O | –OH |
| 4-O-acetyl-NIV (4-ANIV) | –OH | –OAc | –OH | =O | –OH |
| 4-deoxy-nivalenol (DON) | –OH | –H | –OH | =O | –OH |
| 3-O-acetyl-DON (3-ADON) | –OAc | –H | –OH | =O | –OH |
| 15-O-acetyl-DON (15-ADON) | –OH | –H | –OH | =O | –OAc |
| Trichothecin | –H | –OIsoval | –H | =O | –H |

## Trichothecene Biosynthesis in *Fusarium*

The genes related to the synthesis of secondary metabolites are located near each other to form a gene cluster in fungi (Hohn et al. 1993). The *Tri5* gene cluster is responsible for the production of trichothecenes in *Fusarium* spp. (Lee et al. 2002; Kimura et al. 2007). Although gene variations affect the cluster size and the production of the final mycotoxin type, the determined size for *F. graminearum* becomes 27 kb.[1] Partial deletions and/or insertions, complete deletion of the gene, or pseudogene organization are the mechanisms responsible for size differences (Chandler et al. 2003; Brown et al. 2004). The cluster is positioned on chromosome 2 and has 12 *tri* genes consisting of *tri5* (encoding the first enzyme of the biosynthesis pathway) and the other 11 genes responsible for regulatory, biosynthetic, and promoter functions (Fig. 1). Also, four more pathway genes (*tri1*, *tri16, tri101*, and *tri15*) are placed outside of the *Tri5* cluster (Alexander et al. 2004; Seong et al. 2009; McCormick et al. 2011).

In trichothecene biosynthesis, regardless of which type of trichothecene will be produced, *tri4* and *tri5* genes encode enzymes responsible for the initiation of biosynthesis. After these enzymatic steps, the product, isotrichotriol, undergoes two nonenzymatic isomerization steps. The base trichothecene skeleton structure isotrichodermol is formed and presented for further enzymatic processes to the production of different trichothecene types (McCormick et al. 1990). The biosynthetic pathway for trichothecene production has still been examined at the gene expression level. According to the current knowledge, the synthesis of different types of trichothecenes is shown in Fig. 2.

The genes located on the *Tri5* gene cluster have been used in fungal taxonomy and chemotype differentiation assays (Chandler et al. 2003; Kimura et al. 2003). *tri5, tri1, tri101*, and *tri16* genes have been targeted in multi-loci genotyping assays for differentiating the members of the *F. graminearum* species complex (O'Donnell et al. 2008; Yli-Mattila et al. 2009; Liang et al. 2015). *tri5* gene was first used to identify trichothecene-producing *Fusarium* spp. (Hue et al. 1999). Insertion and deletion in *tri7* and *tri13* were also targeted to differentiate DON and NIV-producing *F. graminearum* and *F. culmorum* isolates (Lee et al. 2002; Chandler et al. 2003; Brown et al. 2004). *tri3* and *tri12* were then used to differentiate trichothecenes' sub-chemotypes (Jennings et al. 2004; Wang et al. 2008; Nielsen et al. 2012). In

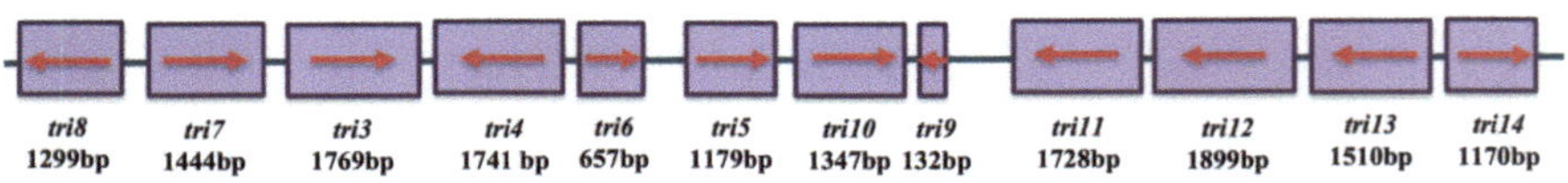


**Fig. 1** Genes are responsible for trichothecene biosynthesis in the *Tri5* gene cluster. The direction of the rows indicates the promoter position, and the row sizes correspond to the actual gene sizes. (Kimura et al. 2003)

[1] https://www.ncbi.nlm.nih.gov/nuccore/20475390/ [Retrieved September 2025]

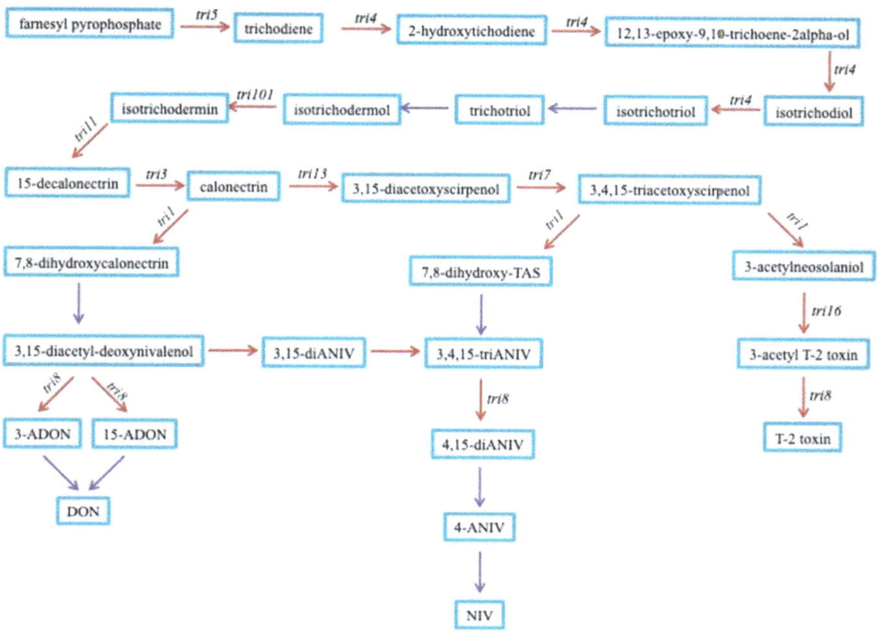


**Fig. 2** Trichothecene biosynthetic pathway. Red arrows indicate enzymatic steps, whereas purple arrows show steps without any defined gene. (Kimura et al. 2003)

vitro and *in planta* chemotype distribution could be easily determined by *tri* genes amplification in a short time.

## Cellular Targets of A- and B- Trichothecenes

Serious mycotoxicoses are associated with the ingestion of moldy products contaminated by trichothecenes. It is crucial to enlighten the cellular processes of trichothecenes for understanding the pathogenesis of different organisms. Investigation of trichothecenes' toxicity mechanisms is an interminable research topic, since they expose various toxic effects in mammals, including feed refusal, vomiting, growth retardation, immunosuppression, and reproductive disorders. These toxic effects indicate the cytotoxicity impact of trichothecenes at cellular and molecular levels (Wang et al. 2021). The low molecular weight property and amphipathic nature of trichothecenes let them easily absorbed through an organism's gastrointestinal membranes and distributed to various tissues and organs (Janik et al. 2021).

Trichothecene accumulation arises ribotoxic stress responses in those cells, which is continued with the activation of MAPKs as apoptotic signals for cells (Shifrin and Anderson 1999; Arunachalam and Doohan 2013). Induced apoptosis in lymphoid, hematopoietic, gastrointestinal, and also neural cells by trichothecenes gives rise to leukopenia, vomiting, diarrhea, muscular weakness, and ataxia (Haschek and Beasley 2009). Also, trichothecenes can decrease immune system activity by suppressing the response mechanisms of T-dependent antigens (Arunachalam and Doohan 2013).

These findings were collected via various elaborate research conducted with both experimental subjects and cell lines. Tissue-level studies, such as investigation of the effect of T-2 toxin on the blood-brain barrier were carried out in mice (Ravindran 2013) and molecular researches were pursued with different cell lines like human neuroblastoma cells IMR-32 (Agrawal et al. 2015), hepatocyte L02 cells (Dai et al. 2016), RAW264.7 murine macrophages (Wu et al. 2014). Also, different yeast strains such as *Saccharomyces cerevisiae* BY4743 and W303 were used as model organisms especially for understanding the toxins' mechanism on eukaryotic ribosomes (McLaughlin et al. 2009; Bin-Umer et al. 2011).

The investigated pathways include inhibition of eukaryotic protein synthesis, cell division, and DNA replication; mitogen-activated protein kinases (MAPKs) signalling pathways activation, membrane integrity alteration, mitochondria function disruption, cytokine gene expression, and cell apoptosis induction. Trichothecenes mostly target actively dividing cells in the thymus, bone marrow, intestinal mucosa, lymph nodes, and spleen with the ability of protein synthesis inhibition (Haschek and Beasley 2009). As the existing literature and models, attacking this constitutive mechanism becomes trichothecenes' primary and most destructive mode of toxic action. It is achieved by preventing the peptide bond formation via the noncovalent binding of trichothecenes to the peptidyl transferase site of 60S ribosomal unit on ribosomes (Feinberg and McLaughlin 1989).

Since T-2 is the most toxic trichothecene, there are many studies on the cellular effect of it in the literature. It was identified in chicken liver cells that T-2 toxin accumulation has resulted in pathological changes, apoptosis, and autophagy in hepatocytes with an increased dose-dependent manner (Janik et al. 2021). Hepatocyte edema, increased volume, and granular cytoplasm was observed as cellular changes. It was observed that the production of reactive oxygen species and cytochrome c translocation between mitochondria and cytoplasm were concluded with T-2-induced mitochondria-mediated apoptosis in hepatocytes. Additionally, increases in the expression of autophagy-related proteins, Beclin-1, ATG5, ATG7, and the LC3-II and autophagy signaling pathway phosphoinositide 3-kinase (PI3K)/ protein kinase B (AKT)/mammalian target of rapamycin (mTOR) were identified in the same study (Yin et al. 2020).

The effects of T-2 toxin on kidney cells were investigated by feeding juvenile goats and mice with a T-2 toxin-contaminated diet. Researchers encountered with certain morphological changes at the cellular level such as pleomorphic rounded forms of mitochondria arising from cristae loss, heterochromatin condensation, indistinct nuclear membrane and degeneration of the epithelial lining of tubules, presence of karyomegaly and binucleation. In addition, apoptosis and necrosis were also determined in kidney epithelial and renal cells (Nayakwadi et al. 2020; Rahman et al. 2021).

Immunological response to T-2 toxin accumulation is another in-demand research topic, and studies were maintained with different cell lines, animal models, and in vitro models (Janik et al. 2021). Minervini et al. (2005) observed early apoptosis via activation of caspase-3, cell membrane damage with their study on two lymphoid human cell lines, MOLT-4 and IM-9. Moreover, it was established that human monocytes' differentiation into macrophages and dendritic cells as the normal immunological response was disturbed via T-2 toxin. Downregulation of CD71 and CD1a, certain cell markers for macrophages and dendritic cells, and upregulation of CD14 (specific monocyte marker) were observed and concluded as T-2 cytotoxicity on monocyte differentiation (Hymery et al. 2009). Rahman et al. (2021) studied with rats and observed humoral and cellular immunity suppression caused by T-2 toxicity. They observed decreases in serum immunoglobulin levels, the number of CD4+ and CD8+, and mRNA expression levels of interleukins. Seeboth et al. (2012) developed an in vitro model of primary porcine alveolar macrophages to demonstrate the effects of T-2 on the activation of macrophages via Toll-like receptors (TLR). They showed a decrease in TLR activation, thereby the decrease in pattern recognition of pathogens, resulting in disruption in the initiation of inflammatory immune responses against viruses and bacteria.

Moreover, T-2 toxin-induced apoptosis was investigated also in neural cells. Different cell lines like human neuroblastoma IMR-32, normal human astrocytes NHA, and mouse neuroblastoma2a N2a were used for that purpose. Researchers established reactive oxygen species (ROS) production, mitochondrial membrane permeability loss, caspase-3 activation, p53 activation, nuclear fragmentation, MAPK signal transduction pathway activation, and apoptosis after T-2 toxin exposure (Weidner et al. 2013; Agrawal et al. 2015; Zhang et al. 2018).

It is also detected that T-2 toxin exposure has a variety of negative effects on fertility. Studies conducted with male and female mice showed that significant decrease in the number of live spermatozoa, a notable increase in abnormal spermatozoa, and a remarkable decrease in spermatozoa with integrated acrosome, low pregnancy rate, and high fetal resorption rate and disruption of estrogen and progesterone the synthesis caused by germ cell apoptosis and oxidative stress (Yang et al. 2010; Shen et al. 2019; Yang et al. 2019; Perveen et al. 2020).

T-2 toxin has major dermal effects like skin fibroblast cell destruction, skin damage, and inflammation, similar to radiation injuries. Studies with local application of toxin had results as degenerative alterations such as vacuolation, ballooning of basal keratinocytes, and infiltration of inflammatory cells in the dermis caused by increased ROS generation, lipid peroxidation, neutrophil-mediated myeloperoxidase activity, inflammatory cytokines and p38 MAPKs levels which all indicate apoptosis in dermal tissue (Pang et al. 1987).

## Clinical Aspect of Mycotoxin Accumulation

The toxic effects of trichothecenes drew attention for the first time in the early years of the Second World War, in Russia (at that time, Soviet Union). Many people suffered from a devastating disease with serious symptoms like necrosis, hemorrhage, and central nervous system effects, often resulting in death. After elaborative research and studies, it is propounded that *F. sporotrichioides* and *F. poae* on harvested grains were responsible for this lethal disease and the name of the disease was determined as alimentary toxic aleukia, ATA (Richard 2007). According to sustained research, it is claimed that trichothecene mycotoxins were the causative agent of this mortal disease.

One of the other diseases is an endemic form of osteoarthritis named as Kashin-Beck disease. The prognosis of the disease occurs in a chronically disabling, deforming, and dystrophic form, on the peripheral joints and spine. The disease is endemic and geographically distributed in a narrow zone in China, Eastern Siberia, and North Korea. Kashin-Beck disease affects preschool or school-level children with the age between 5 and 15. Initially, there are no symptoms related to the disease, and the prognosis begins slowly. However, in the later stages of the disease, shortening of the long bones, thickening and subsequent deformity of the joints, flexor contractures, and muscular atrophy develop. According to the epidemiological data, the responsible agent for the disease is an A-type trichothecene: T-2 toxin (Nelson et al. 1994).

Consumption of scabby grains is also a causative agent of human mycotoxicosis via DON, NIV, fusarenon-X, diacetylnivalenol, and T-2 toxins. Feed refusal, anorexia, nausea, vomiting, headache, abdominal pain, diarrhea, chills, giddiness, and convulsions are the general symptoms of scabby grain mycotoxicosis. Also, T-2 toxin and diacetoxyscirpenol may cause hemorrhagic syndrome for humans and animals, characterized by bloody diarrhea, necrotic oral lesions, gastroenteritis, and hemorrhages in different organs (Nelson et al. 1994).

In addition to acute diseases, if the organism is exposed to a high level of mycotoxin, a toxic response happens rapidly and it may have the risk of sudden death. Also, chronic diseases may arise in the organism according to the low-dose and exposures to trichothecenes, the effect of mycotoxins on organisms' health emerges over a long period, such as cancer (Cimbalo et al. 2020). Therefore, certain international regulatory bodies have carried out risk assessment and regulations for managing mycotoxins' presence in foods and feeds, such as the WHO/FAO Joint Expert Committee on Food Additives (JECFA) and the European Union (EU).

## Clinical Aspect of *Fusarium* Exposure

Fusariosis is the general name of human infections caused by *Fusarium* spp. (Sáenz et al. 2020). Tissue breakdown is the major risk for contamination with organism. Additionally, airborne distribution of pathogens and contamination of water distribution system pieces such as drains, faucet aerators, and shower heads become risk factors especially for hospitals, currently (Sáenz et al. 2020). Also, this organism can adhere to prosthetic body materials such as contact lenses and catheters. In addition to organism intake, presence of mycotoxins enables humoral and cellular immunity suppression that give rise to more severe infection. Keratitis, onychomycosis, peritonitis, and cellulitis are the main types of *Fusarium* infections (Dignani and Anaissie 2004).

Fungal keratitis is a *Fusarium* disease. It leads to corneal damage and visual loss in humans in developing countries. The disease tends to have a higher incidence in the tropical and subtropical regions than in the temperate areas because of the hot, humid climate and an agriculture-based occupation. It is caused by nonspecific ocular injuries from leaves, paddy grain, cow tail, tree branch, and metal pieces (Gopinathan et al. 2002). Keratitis patients are cured with a combination of topical antifungals among topical or systemic antifungal agents, which is effective against *Fusarium*. However, patients late for medical advice are more likely to undergo corneal transplantation since the needed medicinal dosage becomes too high (Dignani and Anaissie 2004).

Furthermore, the wear of contact lenses has the risk of contamination with *Fusarium* spp., especially during improper care. Though, during windy conditions, *Fusarium* spp. may also contaminate the eyes and lenses during use without any inappropriate conditions. Since contact lenses have a soft matrix and high water content, they serve as a convenient surface for rapidly increasing microbial growth. Treatment often requires topical use of natamycin after removing the lenses, but surgery may be required in refractory patients (Simmons et al. 1986; Dignani and Anaissie 2004).

There is a wide distribution of skin-related fusariosis originated on the epidermis. Onychomycosis is one of them, caused by the invasion of *Fusarium* spp. into mostly the great toenails. The fungus is reported as the causative agent of 9–44% of the nondermatophytic molds related nail invasions. Soil contamination is observed in patients with a history of walking with open sandals or barefoot. Cutaneous infections by *Fusarium* spp. may develop on the skin with excessive moisture or trauma observed with skin breakdown. Different types of skin lesions, including granulomas, ulcers, nodules, mycetomas, panniculitis, and necrosis, may be observed during the prognosis. Subcutaneous nodules, ecthyma-like lesions, bullae, or target lesions may develop. The evolution of lesions is a rapid process, usually over a few days like 1–5 days. Moreover, different stages of lesion evolution, such as a combination of papules, nodules, and necrotic lesions, are observed in most patients. Even if there is a variation of skin lesions for disease prognosis, skin lesion patterns may not be associated with any particular *Fusarium* species (Dignani and Anaissie 2004).

Patients with prosthetic materials have the risk for *Fusarium* contamination via plugging or invading of fungi to catheter from the airified points of those materials. In addition, peritonitis and catheter-associated fungemia may occur after certain surgeries during catheter existence. In peritonitis cases, fever, abdominal pain, and decreasing drainage from the peritoneal catheter are observed; however, the clinical presentation may be identified as insidious because of the nonspecific symptoms. Both diseases are treated via catheter removal and systemic antifungal therapy (Raad and Hachem 1995; Bibashi et al. 2002; Dignani and Anaissie 2004).

## Manipulations to the Downregulation of Trichothecene Production

Currently, studies about *Fusarium* spp. and the struggle with diseases caused by this pathogen and their produced mycotoxins are still limited. Therefore, the related researches become mostly to develop pathogen-resistant crops. However, these researches have some constraints. Having too many plant species is the first obstacle for this research area. Moreover, even if the resistant species can be attained after the time- and money-consuming experimental period, their effectiveness and usefulness become low regarding plant production on the field. Studies with antagonistic microorganisms become another type of attempted fungal control method against *Fusarium*. Lastly, treatment with fungicides is also a promising approach. However, resistance mechanisms against these treatments may arise.

The extensive use of fungicides may evoke hazards for health and the environment, and it cannot be guaranteed for an effective decrease of mycotoxins in food and feedstocks. So, up to date, there is no promising and effective solution to *Fusarium* diseases. Therefore, the current situation makes this field an attractive research area (Bilska et al. 2018).

Manipulating growth conditions is a promising approach to develop inhibitory conditions on fungal growth and reduce mycotoxin production. Different research groups have been focused on the investigation of external factor differences (such as pH and temperature), the effect of different organic and inorganic compounds (such as various carbon and nitrogen sources, $Mg^{2+}$, catalase, and $H_2O_2$), and different metabolites (such as transcription factor FgCrz1A, ferulic acid, sinapic acid, triazole derivatives, and different essential oils like camphor and kaempferol) which are revealed as artificial addition of those substances into growth media become effective for inhibition of fungal growth and trichothecene production (Ponts et al. 2007; Jiao et al. 2008; Pinson-Gadais et al. 2008; Boutigny et al. 2009; Gardiner et al. 2009; Kulik et al. 2017; Yörük et al. 2017; Gazdagli et al. 2018; Chen et al. 2019b). As another approach, gene silencing studies are hot topics to regulate biosynthetic pathways. There are many studies for that purpose with different mechanisms. McDonald et al. (2005) and Scherm et al. (2011) worked on certain isolates for silencing *tri6* by RNAi mechanisms. Also, Yörük and Albayrak (2019) focused on silencing *tri4* and *tri5* through siRNA.

Having knowledge about the *Tri5* gene cluster may help to determine the proposed mycotoxin limits in food and feedstocks, to evaluate the infection grade of crops, to quantify the fungal biomass in the field, and to create routine monitoring with the purpose of disease control in the field (Yörük and Albayrak 2014). Also, since trichothecenes can inhibit protein synthesis in other eukaryotic organisms and it tends to maintain their structure even under extreme conditions (e.g., high temperature) (Lauren and Smith 2001; Gutleb et al. 2002), it is impossible to eliminate trichothecenes during food manufacturing and processing (Hazel and Patel 2004). Therefore, having information about gene regulation becomes an obligation for trichothecene detection and quantification on grains and limiting the biosynthesis by fungus cultivation during the crop regarding food safety and health (Boutigny et al. 2009; Yörük et al. 2017). Also, clinical researches on cellular and organism scales and epidemiological data of fusariosis become crucial for rapid characterization of diseases and effective treatment strategies. To correct recognition of the effects of *Fusarium* spp. and trichothecenes, these studies should be conducted across humans, animals, and plants, in other words, the One Health approach[2] (Sáenz et al. 2020).

## References

Agrawal M, Bhaskar ASB, Rao PVL (2015) Involvement of mitogen-activated protein kinase pathway in T-2 toxin-induced cell cycle alteration and apoptosis in human neuroblastoma cells. Mol Neurobiol 51(3):1379–1394

Alexander NJ, McCormick SP, Larson TM, Jurgenson JE (2004) Expression of *tri15* in *Fusarium sporotrichioides*. Curr Genet 45(3):157–162

Arunachalam C, Doohan FM (2013) Trichothecene toxicity in eukaryotes: cellular and molecular mechanisms in plants and animals. Toxicol Lett 217(2):149–158

Basallote-Ureba MJ, Vela-Delgado M, Capote N, Melero-Vara JM, López-Herrera CJ, Prados-Ligero AM, Talavera-Rubia MF (2016) Control of Fusarium wilt of carnation using organic amendments combined with soil solarization, and report of associated *Fusarium* species in southern Spain. Crop Prot 89:184–192

Berthiller F, Crews C, Dall'Asta C, Saeger SD, Haesaert G, Karlovsky P, Oswald IP, Seefelder W, Speijers G, Stroka J (2013) Masked mycotoxins: a review. Mol Nutr Food Res 57(1):165–186

Bibashi E, Kokolina E, Sigler L, Sofianou D, Tsakiris D, Visvardis G, Memmos D (2002) Three cases of uncommon fungal peritonitis in patients undergoing peritoneal dialysis. Perit Dial Int 22(4):523–525

Bilska K, Stuper-Szablewska K, Kulik T, Buśko M, Załuski D, Perkowski J (2018) Resistance-related L-pyroglutamic acid affects the biosynthesis of trichothecenes and phenylpropanoids by *F. graminearum* sensu stricto. Toxins 10(12):492

[2] http://www.onehealthinitiative.com [Retrieved September, 2025]

Bin-Umer MA, McLaughlin JE, Basu D, McCormick S, Tumer NE (2011) Trichothecene mycotoxins inhibit mitochondrial translation--implication for the mechanism of toxicity. Toxins (Basel) 3(12):1484–1501

Boutigny AL, Barreau C, Atanasova-Penichon V, Verdal-Bonnin MN, Pinson-Gadais L, Richard-Forget F (2009) Ferulic acid, an efficient inhibitor of type B trichothecene biosynthesis and *tri* gene expression in *Fusarium* liquid cultures. Mycol Res 113(6–7):746–753

Brown DW, Dyer RB, McCormick SP, Kendra DF, Plattner RD (2004) Functional demarcation of the *Fusarium* core trichothecene gene cluster. Fungal Genet Biol 41(4):454–462

Chandler EA, Simpson DR, Thomsett MA, Nicholson P (2003) Development of PCR assays to *tri7* and *tri13* trichothecene biosynthetic genes, and characterisation of chemotypes of *Fusarium graminearum*, *Fusarium culmorum* and *Fusarium cerealis*. Physiol Mol Biol Plants 62(6):355–367

Chen Y, Kistler HC, Ma Z (2019a) *Fusarium graminearum* Trichothecene mycotoxins: biosynthesis, regulation and management. Annu Rev Phytopathol 57:15–39

Chen L, Tong Q, Zhang C, Ding K (2019b) The transcription factor FgCrz1A is essential for fungal development, virulence, deoxynivalenol biosynthesis and stress responses in *Fusarium graminearum*. Curr Genet 65(1):153–166

Cimbalo A, Alonso-Garrido M, Font G, Manyes L (2020) Toxicity of mycotoxins in vivo on vertebrate organisms: a review. Food Chem Toxicol 137:111161

Cundliffe E, Cannon M, Davies J (1974) Mechanism of inhibition of eukaryotic protein synthesis by trichothecene fungal toxins. Proc Natl Acad Sci 71(1):30–34

Dai C, Li D, Gong L, Xiao X, Tang S (2016) Curcumin ameliorates Furazolidone-induced DNA damage and apoptosis in human hepatocyte L02 cells by inhibiting ROS production and mitochondrial pathway. Molecules 21(8):1061

Dignani MC, Anaissie E (2004) Human fusariosis. Clin Microbiol Infect 10:67–75

Feinberg B, McLaughlin CS (1989) Biochemical mechanism of action of trichothecene mycotoxins. In: Beasley VR (ed) Trichothecene Mycotoxicosis: pathophysiologic effects vol I. CRC Press, Miami, pp 27–36

Foroud NA, Eudes F (2009) Trichothecenes in cereal grains. Int J Mol Sci 10(1):147–173

Frisvad JC, Larsen TO, Thrane U, Meijer M, Varga J, Samson RA, Nielsen KF (2011) Fumonisin and ochratoxin production in industrial *Aspergillus niger*strains. PLoS One 6(8):e23496

Gardiner DM, Kazan K, Manners JM (2009) Nutrient profiling reveals potent inducers of trichothecene biosynthesis in *Fusarium graminearum*. Fungal Genet Biol 46(8):604–613

Gazdağlı A, Sefer Ö, Yörük E, Varol Gİ, Teker T, Albayrak G (2018) Investigation of camphor effects on *Fusarium graminearum* and *F. culmorum* at different molecular levels. Pathogens 7(4):90

Gopinathan U, Garg P, Fernandes M, Sharma S, Athmanathan S, Rao GN (2002) The epidemiological features and laboratory results of fungal keratitis: a 10-year review at a referral eye care center in South India. Cornea 21(6):555–559

Gutleb AC, Morrison E, Murk AJ (2002) Cytotoxicity assays for mycotoxins produced by *Fusarium* strains: a review. Environ Toxicol Pharmacol 11(3–4):309–320

Haschek WM, Beasley VR (2009) Trichothecene mycotoxins. In: Gupta RC (ed) Handbook of toxicology of chemical warfare agents. Academic, pp 53–69

Hazel CM, Patel S (2004) Influence of processing on trichothecene levels. Toxicol Lett 153(1):51–59

Hohn TM, McCormick SP, Desjardins AE (1993) Evidence for a gene cluster involving trichothecene-pathway biosynthetic genes in *Fusarium sporotrichioides*. Curr Genet 24(4):291–295

Hue FX, Huerre M, Rouffault MA, de Bievre C (1999) Specific detection of *Fusarium* species in blood and tissues by a PCR technique. J Clin Microbiol 37(8):2434–2438

Hymery N, Léon K, Carpentier FG, Jung JL, Parent-Massin D (2009) T-2 toxin inhibits the differentiation of human monocytes into dendritic cells and macrophages. Toxicol In Vitro 23:509–519

Ismaiel A, Papenbrock J (2015) Mycotoxins: producing fungi and mechanisms of phytotoxicity. Agriculture 5(3):492–437

Janik E, Niemcewicz M, Podogrocki M, Ceremuga M, Stela M, Bijak M (2021) T-2 toxin-the most toxic trichothecene mycotoxin: metabolism, toxicity, and decontamination strategies. Molecules 26(22):6868

Jennings P, Coates ME, Walsh K, Turner JA, Nicholson P (2004) Determination of deoxynivalenol and nivalenol producing chemotypes of *Fusarium graminearum* isolated from wheat crops in England and Wales. Plant Pathol 53(5):643–652

Jiao F, Kawakami A, Nakajima T (2008) Effects of different carbon sources on trichothecene production and *tri* gene expression by *Fusarium graminearumin*liquid culture. FEMS Microbiol Lett 285(2):212–219

Kimura M, Tokai T, O'Donnell K, Ward TJ, Fujimura M, Hamamoto H, Shibata T, Yamaguchi I (2003) The trichothecene biosynthesis gene cluster of *Fusarium graminearum* F15 contains a limited number of essential pathway genes and expressed non-essential genes. FEBS Lett 539:105–110

Kimura M, Tokai T, Takahashi AN, Ohsato S, Fujimura M (2007) Molecular and genetic studies of *Fusarium* trichothecene biosynthesis: pathways, genes, and evolution. Biosci Biotechnol Biochem 71(9):2105–2123

Kulik T, Stuper-Szablewska K, Bilska K, Buśko M, Ostrowska-Kołodziejczak A, Załuski D, Perkowski J (2017) Sinapic acid affects phenolic and trichothecene profiles of *F. culmorum* and *F graminearum* sensu stricto. *Toxins* 9(9):264

LaMondia JA (2015) *Fusarium* wilt of tobacco. Crop Prot 73:73–77

Lauren DR, Smith WA (2001) Stability of the *Fusarium* mycotoxins nivalenol, deoxynivalenol and zearalenone in ground maize under typical cooking environments. Food Addit Contam 18(11):1011–1016

Lee T, Han YK, Kim KH, Yun SH, Lee YW (2002) *Tri13* and *tri7* determine deoxynivalenol- and nivalenol-producing chemotypes of *Gibberella zeae*. Appl Environ Microbiol 68(5):2148–2154

Liang J, Lofgren L, Ma Z, Ward TJ, Kistler HC (2015) Population subdivision of *Fusarium graminearum* from barley and wheat in the upper Midwestern United States at the turn of the century. Phytopathology 105(11):1466–1474

McCormick SP, Taylor SL, Plattner RD, Beremand MN (1990) Bioconversion of possible T-2 toxin precursors by a mutant strain of *Fusarium sporotrichioides*NRRL 3299. Appl Environ Microbiol. 56:702–706

McCormick SP, Stanley AM, Stover NA, Alexander NJ (2011) Trichothecenes: from simple to complex mycotoxins. Toxins. 3(7):802–814

McDonald T, Brown D, Keller NP, Hammond TM (2005) RNA silencing of mycotoxin production in *Aspergillus* and *Fusarium* species. Mol Plant-Microbe Interact 18(6):539–545

McLaughlin JE, Bin-Umer MA, Tortora A, Mendez N, McCormick S, Tumer NE (2009) A genome-wide screen in *Saccharomyces cerevisiae* reveals a critical role for the mitochondria in the toxicity of a trichothecene mycotoxin. PNAS 106(51):21883–21888

Miedaner T, Cumagun CJR, Chakraborty S (2008) Population genetics of three important head blight pathogens *Fusarium graminearum*, *F. pseudograminearum* and *F. culmorum*. J Phytopathol 156(3):129–123

Minervini F, Fornelli F, Lucivero G, Romano C, Visconti A (2005) T-2 toxin immunotoxicity on human B and T lymphoid cell lines. Toxicology 210(1):81–91

Moretti A (2009) Taxonomy of *Fusarium* genus: a continuous fight between lumpers and splitters. Zb Matice srp za prir nauke 117:7–13

Nayakwadi S, Ramu R, Kumar Sharma A, Kumar Gupta V, Kumar RK, Shirahatti V, Basalingappa PS, KM. (2020) Toxicopathological studies on the effects of T-2 mycotoxin and their interaction in juvenile goats. PLoS One 15(3):e0229463

Nelson PE, Dignani MC, Anaissie EJ (1994) Taxonomy, biology, and clinical aspects of *Fusarium* species. Clin Microbiol Rev 7(4):479–504

Nielsen LK, Jensen JD, Rodríguez A, Jørgensen LN, Justesen AF (2012) *Tri12* based quantitative real-time PCR assays reveal the distribution of trichothecene genotypes of *F. graminearum* and *F. culmorum* isolates in Danish small grain cereals. Int J Food Microbiol 157(3):384–392

O'Donnell K, Ward TJ, Aberra D, Kistler HC, Aoki T, Orwig N, Kimura M, Bjørnstad S, Klemsdal SS (2008) Multilocus genotyping and molecular phylogenetics resolve a novel head blight pathogen within the *Fusarium graminearum* species complex from Ethiopia. Fungal Genet Biol 45(11):1514–1522

O'Donnell K, McCormick SP, Busman M, Proctor RH, Ward TJ, Doehring G, Geiser DM, Alberts JF, Rheeder JP (2018) Marasas et al. 1984 "Toxigenic Fusarium species: identity and Mycotoxicology" revisited. Mycologia 27:1–23

Pang VF, Swanson SP, Beasley VR, Buck WB, Haschek WM (1987) The toxicity of T-2 toxin in swine following topical application: I. Clinical signs, pathology, and residue concentrations. Fundam Appl Toxicol 9(1):41–49

Perveen A, Shen J, Ali Kaka N, Li C (2020) Maternal exposure to T-2 toxin affects puberty genes and delays estrus cycle in mice offspring. Animals 10(3):471

Pinson-Gadais L, Richard-Forget F, Frasse P, Barreau C, Cahagnier B, Richard-Molard D, Bakan B (2008) Magnesium represses trichothecene biosynthesis and modulates *tri5*, *tri6*, and *tri12* genes expression in *Fusarium graminearum*. Mycopathologia 165(1):51–59

Ponts N, Pinson-Gadais L, Barreau C, Richard-Forget F, Ouellet T (2007) Exogenous $H_2O_2$ and catalase treatments interfere with *tri* genes expression in liquid cultures of *Fusarium graminearum*. FEBS Lett 581(3):443–447

Proctor RH, McCormick SP, Gutiérrez S (2020) Genetic bases for variation in structure and biological activity of trichothecene toxins produced by diverse fungi. Appl Microbiol Biotechnol 04(12):5185–5199

Raad I, Hachem R (1995) Treatment of central venous catheter-related fungemia due to *Fusarium oxysporum*. Clin Infect Dis 20(3):709–711

Rahman S, Sharma AK, Singh ND, Prawez S (2021) Immunopathological effects of experimental T-2 mycotoxicosis in Wistar rats. Hum Exp Toxicol 40:772–790

Ravindran V (2013) Animal feed safety. Poultry Development FAO, p 76

Richard JL (2007) Some major mycotoxins and their mycotoxicoses--an overview. Int J Food Microbiol 119(1–2):3–10

Sáenz V, Alvarez-Moreno C, Pape PL, Restrepo S, Guarro J, Ramírez AMC (2020) A one health perspective to recognize *Fusarium* as important in clinical practice. J Fungi 6(4):235

Saharan MS, Kumar J, Sharma AK, Nagarajan S (2004) Fusarium head blight (FHB) or head scab of wheat- a review. Proc Indian Natl Sci Acad 70(3):255–268

Scherm B, Orrù M, Balmas V, Spanu F, Azara E, Delogu G, Hammond TM, Keller NP, Migheli Q (2011) Altered trichothecene biosynthesis in *tri6*-silenced transformants of *Fusarium culmorum* influences the severity of crown and foot rot on durum wheat seedlings. Mol Plant Pathol 12(8):759–771

Seeboth J, Solinhac R, Oswald IP, Guzylack-Piriou L (2012) The fungal T-2 toxin alters the activation of primary macrophages induced by TLR-agonists resulting in a decrease of the inflammatory response in the pig. Vet Res 43(1):1–11

Seong KY, Pasquali M, Zhou X, Song J, Hilburn K, McCormick S, Dong Y, Xu JR, Kistler HC (2009) Global gene regulation by *Fusarium* transcription factors *tri6* and *tri10* reveals adaptations for toxin biosynthesis. Mol Microbiol 72(2):354–367

Shen J, Perveen A, Kaka N, Li Z, Dai P, Li C (2019) Maternal exposure to T-2 toxin induces changes in antioxidant system and testosterone synthesis in the testes of mice offspring. Animals 10(1):74

Shifrin VI, Anderson P (1999) Trichothecene mycotoxins trigger a ribotoxic stress response that activates c-jun n-terminal kinase and p38 mitogen-activated protein kinase and induces apoptosis. J Biol Chem 274(20):13985–13992

Simmons RB, Buffington JR, Ward M, Wilson LA, Ahearn DG (1986) Morphology and ultrastructure of fungi in extended-wear soft contact lenses. J Clin Microbiol 24(1):21–25

Sudakin D (2003) Trichothecenes in the environment: relevance to human health. Toxicol Lett 143(2):97–107

Summerell BA (2019) Resolving *Fusarium*: current status of the genus. Annu Rev Phytopathol 57:323–339

Sun JZ, Liu XZ, McKenzie EH, Jeewon R, Liu JKJ, Zhang XL, Zhao Q, Hyde KD (2019) Fungicolous fungi: terminology, diversity, distribution, evolution, and species checklist. Fungal Divers 95(1):337–430

Torbati M, Arzanlou M, da Silva Santos AC (2021) Fungicolous *Fusarium* species: ecology, diversity, isolation, and identification. Curr Microbiol 78(8):2850–2859

Varga E, Wiesenberger G, Hametner C, Ward TJ, Dong Y, Schöfbeck D, McCormick S, Broz K, Stückler R, Schuhmacher R, Krska R, Kistler HC, Berthiller F, Adam G (2015) New tricks of an old enemy: isolates of *Fusarium graminearum* produce a type a trichothecene mycotoxin. Environ Microbiol 17(8):2588–2600

Wang JH, Li HP, Qu B, Zhang JB, Huang T, Chen FF, Liao YC (2008) Development of a generic PCR detection of 3-acetyldeoxy-nivalenol-, 15-acetyldeoxynivalenol- and nivalenol-chemotypes of *Fusarium graminearum* Clade. Int J Mol Sci 9(12):2495–2504

Wang W, Zhu Y, Abraham N, Li XZ, Kimber M, Zhou T (2021) The ribosome-binding mode of Trichothecene mycotoxins rationalizes their structure-activity relationships. Int J Mol Sci 22(4):1604

Weidner M, Lenczyk M, Schwerdt G, Gekle M, Humpf HU (2013) Neurotoxic potential and cellular uptake of T-2 toxin in human astrocytes in primary culture. Chem Res Toxicol 26(3):347–355

Wu Q, Wang X, Wan D, Li J, Yuan Z (2014) Crosstalk of JNK1-STAT3 is critical for RAW264.7 cell survival. Cell Signal 26(12):2951–2960

Yang JY, Zhang YF, Liang AM, Kong XF, Li YX, Ma KW, Jing AH, Feng SY, Qiao XL (2010) Toxic effects of T-2 toxin on reproductive system in male mice. Toxicol Ind Health 26(1):25–31

Yang X, Zhang X, Zhang J, Ji Q, Huang W, Zhang X, Li Y (2019) Spermatogenesis disorder caused by T-2 toxin is associated with germ cell apoptosis mediated by oxidative stress. Environ Pollut 251:372–379

Yin H, Han S, Chen Y, Wang Y, Li D, Zhu Q (2020) T-2 toxin induces oxidative stress, apoptosis and cytoprotective autophagy in chicken hepatocytes. Toxins. 12(2):90

Yli-Mattila T, Gagkaeva T, Ward TJ, Aoki T, Kistler HC, O'Donnell K (2009) A novel Asian clade within the *Fusarium graminearum* species complex includes a newly discovered cereal head blight pathogen from the Russian Far East. Mycologia 101(6):841–852

Yörük E, Albayrak G (2014) *Tri4* and *tri5* gene expression analysis in *Fusarium graminearum* and *F. culmorum* isolates by qPCR. *The*. Plant Pathol J 13(2):133–138

Yörük E, Albayrak G (2019) siRNA quelling of *tri4* and *tri5* genes related to deoxynivalenol synthesis in *Fusarium graminearum* and *F. culmorum*. J Environ Biol 40(3):370–376

Yörük E, Gazdağli A, Albayrak G (2017) Effects of growth conditions on *tri4* gene expression in *Fusarium culmorum*. J Agric Fac Gaziosmanpaşa Univ 34(2):91–97

Zhang X, Wang Y, Velkov T, Tang S, Dai C (2018) T-2 toxin-induced toxicity in neuroblastoma-2a cells involves the generation of reactive oxygen, mitochondrial dysfunction and inhibition of Nrf2/HO-1 pathway. Food Chem Toxicol 114:88–97

# Clinical Aspects of Fungal Infections

**Erico S. Loreto, Juliana S. M. Tondolo, and Regis A. Zanette**

## Introduction

Fungi can cause a variety of infections in humans, ranging from localized diseases that affect the cutaneous or subcutaneous tissue (e.g., dermatophytosis, onychomycosis, keratitis, eumycetomas, hyalohyphomycosis, and phaeohyphomycosis) to severe and invasive infections and are a significant cause of morbidity and mortality, particularly in hospitalized patients and the immunocompromised population (Richardson 2005; Casadevall 2018; Seagle et al. 2021).

The expanding patient populations at risk of fungal infections are associated with several healthcare-related factors, including the use of more aggressive treatment modalities, such as hematopoietic stem cell transplantation and solid organ transplantation, as well as prolonged corticosteroid therapy and chemotherapy for cancer patients (Casadevall 2018; Webb et al. 2018; Vallabhaneni et al. 2016). More recently, viral infections, such as the seasonal influenza epidemic and the SARS-CoV-2 (COVID-19) pandemic, have made more critically ill patients susceptible to secondary fungal infections (Mina et al. 2022; Rijnders et al. 2020).

Studies that estimate the global frequency of fungal diseases describe that more than 300 million people are affected yearly by severe fungal disorders. The estimated mortality exceeds 1.5 million deaths annually (Bongomin et al. 2017; Richardson and Lass-Florl 2008; Vallabhaneni et al. 2016). The spectrum of fungi as agents of human disease have increased over the past decades (Bongomin et al. 2017), as well as the incidence of multidrug-resistant fungi (Forsberg et al. 2019; Perlin et al. 2017; Garvey and Rowan 2023).

E. S. Loreto (✉) · J. S. M. Tondolo
Sobresp Faculty of Health Sciences, Santa Maria, Brazil

R. A. Zanette
Graduation Program in Biological Sciences: Pharmacology and Therapeutics, Universidade Federal do Rio Grande do Sul (UFRGS), Porto Alegre, Brazil



A. Gomes Rodrigues, A. Gupta (eds.), *Fungal Infections*, Fungal Biology,
https://doi.org/10.1007/978-3-032-06014-3_6

**Table 1** WHO fungal priority pathogens list (WHO 2022)

| Critical group | High group | Medium group |
|---|---|---|
| *Cryptococcus neoformans* | *Nakaseomyces glabrata* (*Candida glabrata*) | *Scedosporium* spp. |
| *Candida auris* | *Histoplasma* spp. | *Lomentospora prolificans* |
| *Aspergillus fumigatus* | Eumycetoma causative agents | *Coccidioides* spp. |
| *Candida albicans* | Mucorales | *Pichia kudriavzeveii* (*Candida krusei*) |
| | *Fusarium* spp. | *Cryptococcus gattii* |
| | *Candida tropicalis* | *Talaromyces marneffei* |
| | *Candida parapsilosis* | *Pneumocystis jirovecii* |
| | | *Paracoccidioides* spp. |

In 2022, The World Health Organization (WHO) published a fungal priority pathogens list to define the species with significant public health impact (Table 1) based on the antifungal resistance, mortality, evidence-based treatment, access to diagnostics, annual incidence and complications, and sequelae (WHO 2022). This chapter aims to overview the recent updates on the epidemiology of these significant fungal pathogens (Table 1).

## Candidiasis

### *Etiology*

More than 300 species of *Candida* are described, but a small percentage are frequent pathogens for humans. *Candida albicans*, *Candida dubliniensis*, *Candida glabrata*, *Candida guilliermondii*, *Candida kefyr*, *Candida krusei*, *Candida lusitaniae*, *Candida parapsilosis*, and *Candida tropicalis* are the most frequent species related to diseases in humans (Takashima and Sugita 2022; Quindos et al. 2018).

*C. albicans*, in most studies, remains the predominant species (Dadar et al. 2018). However, a shift in the etiology was observed in different regions of the world (Quindos et al. 2018; Sharma and Chakrabarti 2023). For example, in Latin America and some Asian countries, *C. parapsilosis* and *C. tropicalis* are recovered as the most common species; in the United States and northwestern Europe, *Nakaseomyces glabrata* (*C. glabrata*) is more frequently observed (Quindos et al. 2018; Ruhnke 2019; Pappas et al. 2018; Lockhart et al. 2017). Of notable concern, *Candida auris* emerged as a multidrug-resistant pathogen and a severe global health threat. First documented in 2009, it has caused several outbreaks in different world regions (Desoubeaux et al. 2022; Geremia et al. 2023).

Several other species of *Candida* have been identified as pathogenic for humans, including *Candida africana*, *Candida blankie*, *Candida bracarensis*, *Candida famata*, *Candida lipolytica*, *Candida nivariensis*, *Candida pulcherrima*, and *Candida rugosa* (Kumar et al. 2022). The increasing number of non-*C. albicans* were associated with species reclassification as the number of immunocompromised patients increased. For example, *C. parapsilosis* was divided into three species (*C. parapsilosis*, *C. metapsilosis*, and *C. orthopsilosis*); *C. auris*, *C. bracarensis*, and *C. nivariensis* are new species described; *C. rugosa*, *Candida sake*, and *Candida zeylanoides* are environmental species that now are associated to human disease (Takashima and Sugita 2022).

Phylogenetical analyses described that *Candida* could be split into more than ten clades. Several reclassifications are underway. For example, *C. guilliermondii*, *C. kefyr*, *C. krusei*, and *C. lusitaniae* were reclassified as *Meyerozyma guilliermondii*, *Dkuveromyces marxianus*, *Pichia kudriavzevii*, and *Clavispora lusitaniae*, respectively; *Lodderomyces/C. albicans* clade harbor *C. albicans*, *C. dubliniensis*, *C. parapsilosis*, and *C. tropicalis* (monophyletic species); *C. auris*, *Candida haemulonii*, *Candida pseudohaemulonii*, and *Candida doubushemulonis* are closed species that may be allocated to the genus *Clavispora*; *Candida bracarensis*, *Candida castelli*, *C. glabrata*, *Candida kungkrabaensis C. nivariensis*, *Candida uthaithanina* belongs to the *Nakaseomyces* clade (Takashima and Sugita 2022; Kidd et al. 2023). The clinical relevance of several reclassifications is unclear, and the denomination of cryptic species as a "complex" is, in most cases, used (Turner and Butler 2014).

## Ecology and Transmission

*Candida* species are ubiquitous yeasts. *C. albicans* is a frequent colonizer of the skin and normal flora of mucocutaneous membranes of humans. Also, it was recovered from soil, hospital environment, food, inanimate objects, animals, and non-animal environments. Non-*C. albicans* species have been recovered from animal and non-animal environments (Ruhnke 2019).

*Candida* infections have an endogenous origin in most cases. The human-to-human transmission was described. A relatively high incidence of carriage on the skin and mucosa of health professional workers was related, reinforcing the evidence that candidiasis can also be an infection acquired from the hospital environment (Lionakis et al. 2020).

## Risk Factors

*Candida* species are opportunistic pathogens. Infections can occur due to characteristics related to the microorganism, the host, or both. The main conditions that predispose human candidiasis are (i) the use of broad-spectrum antibiotics, (ii) mucosal barrier breakdowns, such as those induced by medical interventions or cytotoxic chemotherapy, and (iii) iatrogenic immunosuppression, such as chemotherapy-induced neutropenia or corticosteroid therapy (Pappas et al. 2018). The most common healthcare-associated risks are the intensive care units (ICU) and extended hospital stays (Hohmann et al. 2023; Pfaller and Diekema 2007).

Risk factors for invasive candidiasis often involve abdominal surgery, acute necrotizing pancreatitis, subcutaneous implanted medical devices, critical illness, glucocorticoid use or chemotherapy for cancer, malignant hematologic disease, hemodialysis, immunosuppressive diseases, neonates with low birth weight and preterm infants, presence of a central vascular catheter, total parenteral nutrition, solid organ transplantation, solid organ tumors, and use of broad-spectrum antibiotics (Kullberg and Arendrup 2015; Parslow and Thornton 2022).

## *Incidence*

Several epidemiological characteristics can influence the incidence of *Candida* infections. *Candida* species are among the top four leading pathogen recovery in health-care-associated bloodstream infections, particularly in ICU, and among the top 10 pathogens in other population-based studies. Candidiasis affects around 250,000 individuals and causes more than 50,000 deaths yearly (Kullberg and Arendrup 2015; Wisplinghoff et al. 2004; Magill et al. 2014; Vincent et al. 2009).

For candidemia, unadjusted mortality rates vary widely (from 29 to 76%). The attributed mortality rate ranges from >30 to 40% in the United States, and a median cost of $46,684 was estimated for inpatients (Pappas et al. 2018; Strollo et al. 2017; Wisplinghoff et al. 2004; Cleveland et al. 2012).

The Prospective Antifungal Therapy Alliance (PATH Alliance®) survey reported that *Candida* spp. were the most common fungi isolated (73.4% of 7526 cases) among patients hospitalized with invasive fungal infections. Invasive candidiasis was the primary fungal infection in medical (79.8%, $n = 3713$) and surgical patients (90.7%, $n = 2111$) and those with solid tumors (89.2%, $n = 983$) (Azie et al. 2012).

Among *Candida* spp., non-*C. albicans* (52.2%) were more commonly isolated than *C. albicans* (47.8%), with *N. glabrata* (*C. glabrata*) identified as the second most frequent species (25%). Higher proportions of *P. kudriavzeveii* (*C. krusei*) and C. parapsilosis were reported from hematologic malignancy patients (14.1%, $n = 80$), hematopoietic stem cell transplantation recipients (14.7%, $n = 28$), and neonatal intensive care unit (35.8%, $n = 24$) (Azie et al. 2012).

## *Aspergillosis*

## *Etiology*

The *Aspergillus* genus contains more than 300 species (Samson et al. 2014). *Aspergillus fumigatus* complex members are the most common human pathogens, followed by *A. flavus*, *A. niger*, *A. terreus*, and *A. nidulans* (Thompson 3rd and

Young 2021; Dagenais and Keller 2009; Sabino et al. 2021). Morphologically conventional identification can only define "species sections," such as Fumigati, Flavi, Nidulantes, Usti, and Terrei. Morphologically identical species can only be distinguished by molecular methodologies and are designed as cryptic species and should be reported as "species complex," given the current complexity of *Aspergillus* classification and the lack of sufficient epidemiological data on the new cryptic species.

In the context of the increase in patients with immunosuppression and advances in molecular diagnostics, many cryptic species have been reported recently (Tsang et al. 2020; Fernandez-Pittol et al. 2022). Of concern, cryptic species are frequently more resistant to antifungal agents (Imbert et al. 2021; Sabino et al. 2021; Balajee et al. 2007; Alastruey-Izquierdo et al. 2012).

## Ecology and Transmission

*Aspergillus* species are ubiquitous in the environment and are found in diverse ecological niches, such as soil, plant debris, food, water, animal habitats, and indoor and outdoor air spaces (Latge and Chamilos 2019). Regional environmental characteristics, such as temperature, pH, nutrient conditions, precipitation, and humidity, can drive a fungal adaptation (virulence) and the prevalence of *Aspergillus* species. The correlation of these factors with human infection has been suggested (Panackal et al. 2010; Cadena et al. 2021).

Most human contact with inhaled *Aspergillus* conidia does not cause measurable colonization or disease. The retention of conidia in the lungs can develop various clinical syndromes, from asymptomatic colonization to invasive infection, depending on their immunocompetence status. Uncommon routes of inoculation included ingesting spores via the gastrointestinal tract and direct inoculation via skin injuries (Dagenais and Keller 2009; Challa 2018; Bellmann-Weiler and Bellmann 2019; Thompson 3rd and Young 2021).

## Risk Factors

The major risk factors for aspergillosis included hematopoietic stem cell transplantation (Omrani and Almaghrabi 2017); solid organ transplantation (Silva et al. 2018); hematologic malignancies (van de Peppel et al. 2018); prolonged and profound neutropenia and the use of immunosuppressive drugs, particularly corticosteroid therapy (Li and Denning 2023); treatment with biologic agents (e.g., immunomodulators) (Davis et al. 2020); patients with critical illness and intensive care unit (Kluge et al. 2021); and immunosuppressed patients (Kontoyiannis and Bodey 2002).

Other important predisposing factors are the presence of underline diseases such as severe respiratory viral infections (influenza, respiratory syncytial, or

SARS-CoV-2 virus) (Chen et al. 2020; Verweij et al. 2020; Castro-Fuentes et al. 2022; Nam and Ison 2020), cytomegalovirus infection (Chuleerarux et al. 2021), chronic obstructive pulmonary disease (COPD) (Otu et al. 2023), decompensated cirrhosis (Levesque et al. 2019), liver failure (Falcone et al. 2011), and metabolic comorbidities (e.g., renal disease, diabetes) (Cadena et al. 2021).

## *Incidence*

Data from the Transplant-Associated Infection Surveillance Network (TRANSNET) described that invasive aspergillosis was the most common invasive fungal infection in hematopoietic stem cell transplantation (43%, 425 cases), followed by invasive candidiasis (28%, 276 cases) and mucormycosis (8%, 77 cases). One-year overall mortality reaches 75% (Kontoyiannis et al. 2010).

In the Prospective Antifungal Therapy Alliance (PATH Alliance®) registry, invasive aspergillosis was the second most common pathogen (13.3%, $n$ = 1001), of which *A. fumigatus* (54.2%, $n$ = 543) represented the most common *Aspergillus* species, followed by *A. flavus* (7.4%, $n$ = 74), *A. niger* (6.5%, $n$ = 65), and *A. terreus* (3.2%, $n$ = 32). Invasive aspergillosis was more frequently seen in hematopoietic stem cell transplantation (49.5%, 274 cases) and hematologic malignancy patients (35.2%, 475 cases) (Azie et al. 2012).

# *Cryptococcosis*

## *Etiology*

*Cryptococcus* genus comprises more than 100 species ubiquitously spread in the environment. *Cryptococcus neoformans* and the *Cryptococcus gattii* are the primary etiologic agents of human cryptococcosis. *Cryptococcus albidus* and *Cryptococcus laurentii*, which account for 80% of non-*C. neoformans/gattii* infections, were rarely associated with cryptococcosis in humans (Geddes-Mcalister and Shapiro 2019; Zavala and Baddley 2020; Khawcharoenporn et al. 2007).

Historically, *C. neoformans* and *C. gattii* were distinguished into three varieties (*neoformans, grubii,* and *gatti*), five serotypes (A, D, AD for *C. neoformans*; B and C for *C. gattii*), and more recently, in eight molecular subtypes. However, based on genotyping and phylogenetic studies, it was proposed to split *C. neoformans* into two species (*C. deneoformans* and *C. neoformans*) and *C. gattii* into five species (*C. bacillisporus*, *C. decagatti*, *C. deuterogattii*, *C. gattii*, and *C. tetragattii*) (Hagen et al. 2015). Kwon-Chung et al. (2017) described the recommendation for the use of "*C. neoformans* species complex" and "*C. gattii* species complex" as a practical intermediate step until epidemiological and clinically relevant data on these seven species are available in the literature.

## Ecology and Transmission

*Cryptococcus* spp. are saprobes in nature. *C. neoformans* is globally distributed and isolated in rotting wood, decaying material within hollows of certain tree species, fruit, and soil contaminated by avian manure (such as guano from pigeons). On the other hand, *C. gattii* is limited to tropical and subtropical regions and is classically associated with the eucalyptus and several coniferous trees but not from avian manure. Outbreaks in Canada, Northern USA, and Northern Europe suggest that the ecological niche range of *C. gatti* is not fully recognized. Both species can replicate and survive in environmental scavengers such as nematodes and free-living amoebae (May et al. 2016; Maziarz and Perfect 2016).

Cryptococcosis is acquired primarily by inhalation of aerosolized infectious propagules from the environment (bird guano-enriched areas, soil, or trees). The lungs and the central nervous system are the primary sites of infection, but the eye, skin, and prostate can frequently be involved. Occasionally cases were reported after direct traumatic inoculation. Cryptococcosis can occur in several animal species, but there is infrequent evidence of zoonotic transmission. Human-to-human transmission has not been reported except in contaminated tissue transplants (Williamson et al. 2017; Maziarz and Perfect 2016; Guery et al. 2019; Perfect 2020).

## Risk Factor

Cryptococcosis can occur from apparently immunocompetent patients without evident predisposing factors to severely immunocompromised populations. However, a clear underlying immunocompromised condition was present in most cryptococcosis cases (Diniz-Lima et al. 2022).

The main risk factors for the disease are HIV infection, corticosteroid therapy, idiopathic $CD4^+$ lymphopenia, organ transplantation, malignant and lymphoproliferative disorders, sarcoidosis, monoclonal antibodies treatments (such as adalimumab, alemtuzumab, etanercept, infliximab, or anti-GM CSF), rheumatologic diseases (such as systemic lupus erythematosus and rheumatoid arthritis), chronic liver disease, renal failure and peritoneal dialysis, diabetes mellitus, and hyper-IgM or hyper-IgE syndromes (Maziarz and Perfect 2016; Perfect 2015; Guery et al. 2019; Perfect 2020).

## Incidence

Cryptococcosis was considered an uncommon fungal disease before the 1970s. However, its incidence increased in the AIDS epidemic of the 1980s, rising from 0.8 cases per million individuals in the pre-AIDS era to almost 5 cases per 100,000 persons per year in 1992, during the peak of the AIDS epidemic in the United States. During the 1990s, with the widespread use of fluconazole to treat oral candidiasis and active antiretroviral therapy, the incidence of cryptococcosis declined and stabilized at approximately 1 case per 100,000 persons or less (Hajjeh et al. 1999; van Elden et al. 2000; Perfect 2015; Mcneil and Kan 1995).

*Cryptococcus* species are the most common recovery organisms from invasive fungal infection among non-*Candida* yeasts. They are highly prevalent in HIV-positive patients, followed by solid organ transplantation recipients (Azie et al. 2012). HIV-associated cryptococcosis mortality remains high, particularly in patients in regions with less access to medical care. Globally, 15% of AIDS-related deaths were attributed to cryptococcal meningitis, and cryptococcal-related deaths are estimated at 181,100, with 135,900 cases (75%) occurring in sub-Saharan Africa (Pasquier et al. 2018; Goldberg et al. 2018; Rajasingham et al. 2017).

In HIV-negative patients, cryptococcosis can occur in transplant recipients and other patients with defects in cell-mediated immunity (Williamson et al. 2017). In a multicenter, longitudinal cohort study in the United States, the demographics of 145 HIV-negative patients with cryptococcosis demonstrated that solid organ transplantation (33.8%, 49 cases) was the primary predisposing disease, followed by autoimmune syndromes (15.9%), hematologic malignancy (11.7%), decompensated liver disease (9.7%), solid tumor (5.6%), primary immunodeficiency (2.1%), and hematopoietic stem cell transplantation (2.8%). Glucocorticoid therapy and cytotoxic chemotherapy were the immunosuppressive medications described for more than 40% of patients. Central nervous system involvement was observed in 71 patients (49%) (Marr et al. 2019).

## Mucormycosis

## Etiology

Mucormycosis is caused by fungi from the order Mucorales, which comprises 261 species in 55 genera (Skiada et al. 2020). Gunathilaka et al. (2023) described 46 species as agents of this disease, including 14 species belonging to the genus *Mucor*, 9 to the genus *Rhizopus*, 2 to the genus *Rhizomucor*, 4 to the genus *Lichtheimia* (formerly *Absidia*), 5 to the genus *Apophysomyces*, 3 to the genus *Cunninghamella*, 5 to the genus *Saksenaea*, *Actinomucor elegans*, *Cokeromyces recurvatus*, *Mycotypha microspore*, *Syncephalastrum racemosum*, and *Thamnostylum lucknowense*.

The species prevalence can vary depending on geographical characteristics. *Rhizopus* spp. was the most common agent of mucormycosis (around 48% of cases), followed by *Mucor* (14–18%), *Lichtheimia* (5–13%), *Cunninghamella* (6–7%), *Apophysomyces* (5–7%), *Rhizomucor* (4–6%), *Saksenaea* (2–5%), and uncommon species representing less than 10% of cases (Jeong et al. 2019; Roden et al. 2005).

## *Ecology and Transmission*

The species from Mucorales are ubiquitous fungi found in decaying organic matter, soil, contaminated foods, animal excrement, and compost piles (Steinbrink and Miceli 2021). A higher incidence of mucormycosis can be associated with seasonal variations, and most infections are described in climates with decreased precipitation and higher temperatures (Sivagnanam et al. 2017; Shpitzer et al. 2005). Extreme weather and climate events, like hurricanes, tornados, and post-tsunami, are associated with uncommon species' mucormycosis (Neblett Fanfair et al. 2012; Andresen et al. 2005).

The inhalation of spores from environmental sources is the primary mode of acquisition of mucormycosis. After the adherence of spores to the nasal epithelium, the infection (sinusitis) can manifest as rhino-cerebral involvement, affecting the nasal, oral, ocular, brain, and other associated tissues (Prakash and Kumar 2023). Infection by cutaneous or percutaneous route can occur after penetrating wounds, trauma, direct injection of spores, and burns (Skiada et al. 2022). Gastrointestinal mucormycosis is less common and can occur after ingesting spores from contaminated food or medical devices. This clinical manifestation mainly affects patients with immunodepression, such as premature infants, malnourished patients, or those on dialysis (Didehdar et al. 2022; Spellberg 2012).

## *Risk Factors*

Predisposing factors for mucormycosis include diabetes mellitus, hematological malignancies, hematopoietic stem cell transplantation, solid organ malignancies, solid organ transplantation, therapies with corticosteroids or other immunosuppressive drugs, iron overload, trauma, prolonged neutropenia, malnourishment, neonatal prematurity, illicit intravenous drug use, renal failure, AIDS, and liver diseases, (Skiada et al. 2020; Reid et al. 2020).

Mucormycosis is a rare clinical condition in patients without an apparent predisposing risk factor. In this group of patients, the infection is more common after the direct inoculation of fungus due to trauma, surgery, or burns (Hassan and Voigt 2019). Outbreaks of mucormycosis were described after natural disasters and in the healthcare environment (from infected bandages, laundry, and hospital construction) (Steinbrink and Miceli 2021).

The underlying disease correlates to the infection site. For example, sinusitis and rhinocerebral mucormycosis are associated with diabetes mellitus; trauma commonly results in cutaneous mucormycosis; pulmonary infection is linked with hematological malignancies (Skiada et al. 2020).

## Incidence

The incidence of mucormycosis is increasing and can vary considerably according to the geographical area, population, diagnostic procedures, and analysis periods (Saegeman et al. 2010; Guinea et al. 2017; Bitar et al. 2009; Kontoyiannis et al. 2016):

Mucormycosis reached 1.7 cases per one million individuals in the San Francisco Bay area from 1992 to 1993 (Rees et al. 1998) and 0.12 per 10,000 mucormycosis-related hospitalizations from 2005 to 2014 (Kontoyiannis et al. 2016) in EUA; 0.7 cases per million in 1997 and 1.2 per million in 2006 in France (Lanternier et al. 2012); 0.019 cases/10,000 patient-days in 2000 and 0.148 cases/10,000 patient-days in 2009 in Belgium (Saegeman et al. 2010); 1.2 cases per 100,000 from 1988 to 2006 and 3.2 cases per 100,000 from 2007 to 2015 in Spain (Guinea et al. 2017); 12.9 cases per year from 1990 to 1999, 35.6 cases per year from 2000 to 2004, and 50 cases per year from 2006 to 2007 in India. The estimated disease prevalence is around 70 times higher in India than in global data (Prakash and Chakrabarti 2021; Chakrabarti et al. 2019).

The mortality and morbidity rates depend on the affected organ, Mucorales species, and the patient's medical status. The disease can be highly aggressive, and mortality rates can reach 46% in sinus mucormycosis, 73% in mucormycosis after voriconazole treatment, 76% in pulmonary disease, and 96% in disseminated infections (Roden et al. 2005; Trifilio et al. 2007).

Mucormycosis cases (data from autopsy) represent the third most common cause of invasive fungal infection, after candidiasis and aspergillosis (Dignani 2014). Data from Transplant-Associated Infection Surveillance Network show that mucormycosis was the third most common cause of invasive fungal infection (8%) in hematopoietic stem cell transplantation (Kontoyiannis et al. 2010) and the sixth most common (2%) among organ transplant recipients (Pappas et al. 2010).

Hematopoietic stem cell transplantation and hematologic malignancies are the leading underlying conditions in mucormycosis cases in developed countries. In developing countries, particularly India, the disease's major causes are uncontrolled diabetes, trauma, and, more recently, COVID-19 syndrome (Petrikkos et al. 2012; Challa 2019; Singh et al. 2021).

## Pneumocystosis

## Etiology

*Pneumocystis jirovecii* is the agent of pneumocystosis in humans. Species from the genus *Pneumocystis* have been described in several mammalian species (Weissenbacher-Lang et al. 2023) and are highly specific to the host: *P. jirovecii* in *Homo sapiens* (the only species associated with human infection and not detected in other mammals); *Pneumocystis carinii* and *Pneumocystis wakefieldiae* in *Rattus norvegicus*; *Pneumocystis murina* in *Mus musculus*; and *Pneumocystis oryctologi* in rabbits (Ma et al. 2018; Cisse and Hauser 2018).

## Ecology and Transmission

*Pneumocystis* species have a worldwide distribution and inhabit their hosts' lungs almost exclusively. Since *Pneumocystis* spp. cannot be convincingly cultured in vitro, the best experimental protocols are based on natural or induced host-specific infections, such as *Pneumocystis carinii* infection in laboratory rats. Genetic analysis demonstrated the absence of essential enzymes and biosynthetic pathways that make these microorganisms dependent on acquiring nutrients directly from their hosts (obligate parasites) (Walzer 2013; Weyant et al. 2021; Hauser 2021; Cushion et al. 2021; Ames et al. 2023).

The natural reservoir of *P. jirovecii* remains unknown. Studies of environmental sources have detected *Pneumocystis* DNA in water, air, and soil samples. The first contact with *P. jirovecii* occurs in childhood. Studies of seroconversion evidence asymptomatic acquisition of the fungus at early ages: Vargas et al. (2001) described seroconversion in 85% of healthy infants by 20 months of age; Medrano et al. (2003) found increased seropositivity with age from 6 to 13 years, suggesting a continuous exposure to the *P. jirovecii* during infancy; Similarly, Respaldiza et al. (2004) described an overall seroprevalence of 73% in an age-related increase (52% at 6 years, 66% at 10 years and 80% at 13 years). A lower proportion of seropositivity in the adult population suggests a more effective immune system response after repeated exposure to the microorganism (Medrano et al. 2005; Medrano et al. 2003).

Pneumocystis is transmitted through the respiratory route. However, it is unclear whether there is spread by direct contact or droplet spread (Vera and Rueda 2021). In addition, transplacental transmission (vertical transmission) in humans was supported by a molecular study (Montes-Cano et al. 2009).

The mechanisms by which PJP develops have been discussed from two points of view: (a) reactivation of latent infection or (b) primary infection. The first is the older theory by which *P. jirovecii* becomes part of resident microbial flora during infancy and remains dormant until a decline of the immune response when a PJP can occur after the active replication of the pathogen. On the other hand, the primary infection theory suggests that exposure to *P. jirovecii* is transient after recurrent exposition to the fungus in environmental sources. However, genetic typing data from *P. jirovecii* ribosomal RNA in recurrent PJP episodes demonstrated that reinfection and reactivation could occur (Apostolopoulou and Fishman 2022).

## *Risk Factors*

Gardening, hiking, and other outdoor activities have been demonstrated to be independent risk factors for *P. jirovecii pneumonia* (PJP). Increased ambient sulfur dioxide is an air-pollution factor related to the risk of hospitalization with PJP. However, most studies evaluated the climatologic aspect, and there is an association between increased cases of PJP and the warmer periods of the year. The correlation between specific *P. jirovecii* genotypes and seasonal variation was reported (Walzer 2013; Dohn et al. 2000; Miller et al. 2007; Fishman 2020).

The occurrence of PJP is correlated to severely immunocompromised patients, principally in HIV/AIDS population, and with other immunosuppressed conditions, i.e., autoimmune disorders, cancers, transplantation, and chronic lung disease, especially obstructive pulmonary disease (COPD) or emphysema. The risk of pneumocystosis also increases with corticosteroids or other immunosuppressants (such as cancer chemotherapy, lymphocyte-depleting antibodies, and calcineurin inhibitors), cytomegalovirus infection, age (particularly over 60 years old), exposure to infected individuals, malnutrition, smoking, and alcoholism (Ma et al. 2018; Fishman 2020).

## *Incidence*

The *Pneumocystis* spp. is the leading cause of infection in patients with HIV/AIDS, particularly in regions with less access to highly active antiretroviral combination therapy (HAART). The global prevalence is estimated to be greater than 400,000 cases yearly, while the mortality can vary from 10 to 30% (Armstrong-James et al. 2014; Limper et al. 2017). The incidence tends to be reduced when HAART is available. Still, the disease remains at high rates in HIV-misdiagnosed patients and in those who do not have access to or interrupt HAART (Krajicek et al. 2009; White et al. 2018).

The estimated burden of PJP, based on a review of 133,487 cases published in 40 articles evaluated by Leading International Fungal Education (LIFE), described that 77% of patients ($n$ = 102,955) were reported in Africa, followed by America (10%), Europe (7%), and Asia (6). The global incidence was estimated at 5.79 cases per 100,000. The lowest rate was observed in Bangladesh (0.04 cases per 100,000), and the highest was in Nigeria (48.3 cases per 1000,000). Nigeria ($n$ = 74,595), Kenya ($n$ = 17,000), and Tanzania ($n$ = 9600) had the highest number of cases, while Denmark ($n$ = 2), Hungary (5), Qatar (15), and Israel ($n$ = 26) had the lowest (Bongomin et al. 2017).

## *Histoplasmosis*

## *Etiology*

Histoplasma species were initially classified according to their geographic distribution and clinical manifestations. *Histoplasma capsulatum* var. *capsulatum* is widely distributed, and *Histoplasma capsulatum* var. *duboisii* is found in Africa. Both are the agents of histoplasmosis in humans (Arauz and Papineni 2021). *Histoplasma capsulatum* var. *farciminosum* is a horse pathogen reported mainly in Africa and Asia (Scantlebury et al. 2016). However, more recent phylogenetic studies reclassify *H. capsulatum* into at least 14 phylogenetic groups and four lone lineages, which corroborates the high diversity of the genus *Histoplasma* in the world. Currently, *Histoplasma* is considered a complex of cryptic species (Taylor et al. 2022).

## *Ecology and Transmission*

*Histoplasma* species are ecologically adapted to specific environments. They are commonly found in soil enriched with organic matter, bird guano, or bat droppings, which provide a nutrient-rich substrate for their growth. Consequently, the fungus can commonly be found in areas with large populations of birds or bats, such as caves, chicken coops, and old buildings. The fungus grows in warm, humid environments and can survive long in the soil, especially in areas with a high concentration of organic matter (Teixeira Mde et al. 2016; Diaz 2018).

The primary transmission mechanism of histoplasmosis is the inhalation of conidia or mycelial fragments that were aerosolized due to soil disruption. Rarely the host-adapted yeast form can be transmitted directly through tissue transplantation or a laboratory accident. Human-to-human transmission via the pulmonary route has not been reported (Woods 2016; Arauz and Papineni 2021).

## *Risk Factors*

Several risk factors have been associated with the development of histoplasmosis, including environmental exposure and occupation, immune status, and age. Environmental exposure to *H. capsulatum* is the most significant risk factor for developing the infection. Activities that disturb the soil, such as agricultural activities, construction, excavation, gardening, mining, and spelunking, increase the risk of exposure to the fungus (Valdez et al. 2022).

*H. capsulatum* produces a self-limited, asymptomatic infection in most immunocompetent patients. However, individuals with compromised immune systems, such as those living with HIV/AIDS, organ transplant recipients, and patients receiving immunosuppressive therapy such as steroids and TNF-α inhibitors, are at higher risk of developing the disseminated disease (Kauffman 2008; Wheat et al. 2016; Scully and Baddley 2018).

Elderly individuals are more susceptible to the disease due to a decline in immune function and comorbidities such as chronic obstructive pulmonary disease (COPD) and diabetes mellitus (Specjalski et al. 2019; Barros et al. 2023; Hasmoni et al. 2010).

## Incidence

Histoplasmosis is highly endemic in the American continent. Based on *Histoplasma* antigen detection, the estimated prevalence in the general population of Latin American countries is around 32%. Significant variations in prevalence correlate with differences in clime, soil composition, and patient predisposing factors such as HIV coinfection. For example, the prevalence of histoplasmosis in Mexico ranges from 5% to 50%. It is greater than 40% of patients with HIV in Brazil's Central-Northeast region, in contrast to 10% in southern Brazil (Adenis et al. 2018; Colombo et al. 2011; Falci et al. 2019).

The Mississippi and Ohio River Valleys regions are the most endemic in the United States. *Histoplasma* antigen detection tests revealed an 80% positivity among young adult males from these regions, and incidence was estimated to be 6.1 cases per 100,000. However, outbreaks occur in many areas of the United States (Benedict and Mody 2016; Benedict et al. 2015; Barros et al. 2023).

Outside the Americas, increased prevalence and locations of endemicity have been described in China, India, and South East Asia (Arauz and Papineni 2021; Linder and Kauffman 2019).

# Paracoccidioidomycosis

## Etiology

At present, the genus *Paracoccidioides* harbors seven species. Until 2006, it was thought to contain a single species, *Paracoccidioides brasiliensis*. More recently, molecular and genetic studies of *Paracoccidioides*, initially splitting *P. brasiliensis* into two species, *P. brasiliensis,* and *Paracoccidioides lutzii,* and later revealed that *P. brasiliensis* is a complex of distinct phylogenetic species: *P. brasiliensis* sensu lato, *Paracoccidioides americana*, *Paracoccidioides restrepiensis*, and *Paracoccidioides venezuelensis* (Turissini et al. 2017). *P. brasiliensis* sensu stricto is widely distributed in Latin America and is the leading human agent of paracoccidioidomycosis (Hahn et al. 2022).

In 2021, two uncultivated fungal mammalian pathogens, *Paracoccidioides cetii* and *Paracoccidioides loboi* were included in the genus (Vilela et al. 2021). *P. cetii* is a dolphin pathogen genetically related to species of the *P. brasiliensis* complex, while *P. loboi* (formerly *Lacazia loboi*) is genetically closer to *P. lutzii* (Rodrigues et al. 2023).

## Ecology and Transmission

The ecological niche of *Paracoccidioides* is still not entirely known. Paracoccidioidomycosis directly correlates with agricultural workers, humid environments, and forests, suggesting that the disease develops after repeated contact with the fungus in the soil. Humans and armadillos (*Dasypus novemcinctus*) are the main accidental hosts of *Paracoccidioides* species. However, it has been detected molecularly in dogs and other animals presumably infected in rural and peri-urban regions (Hahn et al. 2022; de Macedo et al. 2020).

Inhalation of fungal propagules is the main proposed route of infection. Clinical and experimental data have ruled out traumatic implantation as a route of infection. This difficulty in determining the exact route of infection is associated with the prolonged latency period of the disease (30 or more years), absence of outbreaks, and sporadic isolation of the fungus from the environment. Human-to-human transmission has not yet been documented. Usually, the primary infection is asymptomatic or has nonspecific symptoms that rarely progress to clinical manifestations (Benard 2021).

## Risk Factors

Paracoccidioidomycosis is endemic in tropical and subtropical regions of Latin America. It is usually associated with forested areas, temperatures between 14 and 20 °C, abundant waterways, annual rainfall between 800 and 2000 mm, and certain types of crops, such as coffee and tobacco (Hahn et al. 2022).

Individuals who work with agricultural activities in regions endemic to paracoccidioidomycosis (e.g., Brazil, Colombia, Venezuela, and Argentina) are at greater risk of acquiring the infection. The disease is more related to men than women, aged between 30 and 50 years. Other predisposing factors include alterations in the immune system (e.g., HIV), cancer, immunosuppressive therapy, tuberculosis, and Chagas' disease. Furthermore, there is a link between alcohol intake and smokers with an increased risk for paracoccidioidomycosis (Mendes et al. 2017; Martinez 2015; Bellissimo-Rodrigues et al. 2011).

## Incidence

Most cases (>80%) were reported in Brazil. Other countries that reported cases of the disease were Argentina, Colombia, Bolivia, Ecuador, Peru, and Venezuela. Few or no cases have been described in Central American countries, Paraguay, Uruguay, Chile, Suriname, Guianas, and Caribbean Islands (Martinez 2017; Pecanha et al. 2022). More than 100 cases of paracoccidioidomycosis have been described among other regions outside Latin America. Patients (immigrants or travelers) had a history of traveling to endemic countries (Pecanha et al. 2022).

An annual incidence of 3–4.3 new cases per million inhabitants in Brazil is estimated. However, in areas of high endemicity, these rates can reach 10–30 new cases per million inhabitants per year (Bellissimo-Rodrigues et al. 2011; Hahn et al. 2022; Coutinho et al. 2015). Coutinho et al. (2002) described a mortality of 1.45 per million inhabitants after analyzing 3181 deaths from paracoccidioidomycosis in Brazil over 16 consecutive years (1980 to 1995). Chronic forms of the disease have high morbidity, and around 50% of patients have sequelae, despite treatment (Pecanha et al. 2022).

# Coccidioidomycosis

## Etiology

*Coccidioides immitis* and *Coccidioides posadasii* are the two species of the *Coccidioides* genus identified as the etiologic agents of coccidioidomycosis. Despite the genetic distinction between the two species, both have similar clinical manifestations and susceptibility to antifungal agents, in addition to a few phenotypic differences. Considering that molecular differentiation of species is not routine in laboratories, many reports in the literature about *C. immitis* may be related to both species. "*Coccidioides* spp." is recommended when species determination is unavailable (Kirkland et al. 2022; Kandemir et al. 2022).

## Ecology and Transmission

The precise ecology of *Coccidioides* species is not entirely understood. It is presumed that the fungus inhabits the soil, particularly in arid and semi-arid regions, and that disease acquisition occurs through direct exposure or airborne dissemination of the fungus, which can cause infection after its inhalation by the host (Ampel 2020).

Despite the apparent preference of the fungus for specific soils (Fisher et al. 2007), its isolation from the environment is notoriously difficult. Taylor and Barker

(2019) proposed that *Coccidioides* is not just an environmental saprophyte but an endozoan capable of surviving in granulomas of small animals that live in the soil, which can serve as an environmental reservoir for the fungus.

## *Risk Factors*

Coccidioidomycosis can occur in people exposed to the fungus in endemic areas. Occupational hazards are related to workers' exposure to dust and soil disturbance, such as soil excavation, construction, archaeology, agriculture, oil and gas extraction, firefighting, and during military maneuvers. In addition, an increased risk of acquiring the disease may occur after natural events, such as earthquakes and dust storms (de Perio et al. 2019; Galgiani 2020).

Immunocompromised patients are highly susceptible to coccidioidomycosis. Predisposing factors for severe disease include HIV/AIDS, transplant patients, and immunosuppressive therapy such as corticosteroids, chemotherapy, and TNF inhibitors. Race/ethnicity and pregnancy are also predisposing factors for the disease (Odio et al. 2017; Williams and Chiller 2022).

## *Incidence*

Coccidioidomycosis is endemic in several regions of the United States, including Arizona, California, New York, Nevada, New Mexico, Utah, Washington, and Texas. Hyperendemic regions comprise southern Arizona and the San Joaquin Valley in California. Areas of endemicity outside the United States include Mexico, Central America (Guatemala, Honduras, and Nicaragua regions), and regions of South America, including Argentina, Bolivia, Brazil, Colombia, Paraguay, and Venezuela (Williams and Chiller 2022; Talamantes et al. 2007).

According to CDC data, there has been a considerable increase in the cases of coccidioidomycosis reported since 2014. After a decline between 2012 and 2014 (17,802 to 8232 cases annually), the rates more than doubled and reached about 20,000 in 2019. Most cases (>95%) were reported in Arizona and California. Between 2014 and 2019, there was an increase from 84.4 to 144.1 and 6 to 22.5 cases per 100,000 population in Arizona and California, respectively. Mortality data indicate that approximately 160 patients die yearly from coccidioidomycosis (CDC 2023; Williams and Chiller 2022; Noble et al. 2016).

## Fusariosis

## Etiology

Current taxonomic classification grouped the more than 300 species of *Fusarium* in more than 20 species complexes, of which around 10 complexes are correlated with disease in humans: *Fusarium solani*, *Fusarium oxysporum*, *Fusarium fujikuroi*, *Fusarium incarnatum-equiseti*, *Fusarium dimerum*, *Fusarium chlamydosporum*, *Fusarium sambucinum*, *Fusarium concolor*, *Fusarium lateritium*, and *Fusarium sporotrichioides*. The most clinically relevant species are listed in *F. solani* (around 40 to 60% of infections), *F. oxysporum* (~20%), and *F. fujikuroi* species complexes. Notably, *Fusarium falciforme* (formerly *Acremonium falciforme*) are moved to the *F. solani* species complex (O'donnell et al. 2013; Al-Hatmi et al. 2016b; Hospenthal 2020b; van Diepeningen et al. 2014).

## Ecology and Transmission

Species in the genus *Fusarium* are ubiquitous fungi commonly found in soil, organic debris, plants, and water. They are a significant cause of economic losses in the agricultural field worldwide and can produce biofilm in hospitals and other water systems. *Fusarium* species are globally distributed, but endemic regions are in tropical and subtropical areas (Al-Hatmi et al. 2016a).

Soft tissue infections, including mycetoma, are associated with traumatic inoculation of the pathogen in the healthy host. The inhalation of spores into the paranasal sinuses and lungs or minor trauma is the suggested mode of acquisition of fusariosis in immunocompromised patients (Hospenthal 2020b).

Other portals of entry are the gastrointestinal tract and indwelling intravascular catheters. Onychomycosis with paronychia is the most frequent preexisting superficial lesion predisposing to disseminated fusariosis. Additionally, contaminated water has been suggested as a source of the disease, either by inhalation of aerosols or direct contact with the damaged skin (Nucci et al. 2021).

## Risk Factors

The *Fusarium* species are associated with localized or invasive infections in immunocompetent and immunosuppressed individuals (Guarro 2013).

Acute myeloid leukemia, prolonged neutropenia, and hematopoietic stem cell transplants are the main predisposing factor for disseminated fusariosis. Other predisposing factors include diabetes mellitus, kidney transplant, immunosuppress treatment, Varicella zoster virus, HIV infection, myeloblastic leukemia, trauma, burns, and heat stroke (Batista et al. 2020; Hospenthal 2020b).

## *Incidence*

The burden of fusariosis has not been established, and data from case reports and clinical case series are used to estimate the course of the disease. Al-Hatmi et al. (2016a), reviewing 388 cases of fusariosis published in the literature between 1958 and 2015, described cases of the disease worldwide, particularly in tropical and subtropical regions, such as Brazil, China, Colombia, India, and Mexico. However, reports of the disease were frequent in Australia, South Africa, the United States, Turkey, and some European countries. Most *Fusarium* infections were superficial and subcutaneous (44.8%, $n = 174$), followed by disseminated (28%, $n = 109$) and deep (2.57%, $n = 10$) infections.

*Fusarium* species represent the third most prevalent species (20.6%, $n = 65$) among filamentous fungi associated with invasive fungal infections among hospitalized patients in North America (Azie et al. 2012).

Data from 233 patients with invasive fusariosis, retrospectively reviewed (1985–2000 and 2001–2011) from 44 diagnostic centers and 11 countries, reported that most patients (92%) had hematological disorders. Acute myeloid leukemia (91 cases) and acute lymphoid leukemia (46 cases) were the most frequent underlying diseases. Fuariosis after hematopoietic cell transplantation was observed in 104 patients (Nucci et al. 2014).

The mortality of fusariosis depends on several factors, such as the infection's severity, the patient's immune status, and underlying medical conditions. Fusariosis can be a severe and life-threatening infection, particularly in people with weakened immune systems. Studies have shown that the mortality rate of invasive fusariosis ranges from 30% to 90%, depending on the patient's underlying conditions (Tortorano et al. 2014; Nucci et al. 2015).

## *Talaromycosis*

## *Etiology*

*Talaromyces marneffei* (formerly *Penicillium marneffei*) is the etiological agent of talaromycosis (formerly penicilliosis). *T. marneffei* is the only thermally dimorphic *Talaromyces* spp. or *Penicillium*-like fungi associated with human systemic mycosis. However, other *Penicillium* spp., including *Penicillium chrysogenum*, *Penicillium cluniae*, *Penicillium digitatum*, *Penicillium notatum*, *Penicillium roqueforti*, and *Penicillium stipitatus* can rarely cause infection in humans and other animals (Wang et al. 2023).

## *Ecology and Transmission*

*T. marneffei* is a saprophytic fungus with a reservoir (ecological niche) in healthy bamboo rats (*Rhizomys pruinosis*, *Rhizomys sinensis*, *Rhizomys sumatrensis*, and *Cannomys badius*), and the soil nearly their burrows in the endemic regions of talaromycosis. However, these reservoirs' role in transmitting pathogenic *T. marneffei* to humans is unclear (Wang et al. 2023; Hospenthal 2020b).

Human infection is associated with inhaling airborne conidia of *T. marneffei* from soil or related environmental sources, particularly in rainy seasons. Yeast phase conidia invade the respiratory tract, are phagocytosed by macrophages, and can spread to other host tissues and organs via blood circulation (Narayanasamy et al. 2021; Xu et al. 2023). Thus, *T. marneffei* is considered a facultative intracellular pathogen. That can be found inside host patients' macrophages and tissue histiocytes (Pruksaphon et al. 2022).

Direct infection via the gastrointestinal tract is rare (Ling et al. 2022). Gupta et al. (2022) reported the first subcutaneous infection in a renal transplant patient. In addition, the laboratory-acquired disease has been described by direct inoculation of the fungus into the cutaneous tissue (Wang et al. 2023).

## *Risk Factors*

Immunosuppression is the leading risk factor associated with talaromycosis. Rarely occurs in immunocompetent patients with other comorbidities, such as chronic obstructive pulmonary disease (COPD) (Yu et al. 2018; Narayanasamy et al. 2021).

The leading risk factor for talaromycosis is advanced HIV disease (CD4 cell count <200 cells/μL). In endemic areas, it is the most common illness among patients living with HIV. *T. marneffei* infections have been increasingly reported in other groups of patients with impaired cell-mediated immunity, including patients with anti-interferon gamma (anti-IFN-g) autoantibodies, systemic lupus erythematosus, cancers patients, hematopoietic stem cell or solid organ transplantation, hematological malignancies, and novel therapies such as anti-CD20 monoclonal antibodies or kinase inhibitors (Wei et al. 2021; Cao et al. 2019; Chan et al. 2015).

The disease occurs primarily in tropical or subtropical regions of Asia, where it is endemic. Also, there is a high association between tropical monsoon weather and talaromycosis. In this period, a 30 to 73% increase in cases was reported in Thailand, Vietnam, and China. There is a greater risk in farmers than nonfarmers of developing talaromycosis, highlighting the association of the disease with the impoverished population. Travelers visiting endemic areas are also potentially vulnerable (Narayanasamy et al. 2021; Wang et al. 2023).

## *Incidence*

In a systematic review and meta-analysis, Qin et al. (2020) described that from 159,064 patients with HIV, there was a pooled prevalence of 3.6% of talaromycosis. The highest prevalence was observed in Vietnam (6.4%), followed by Thailand, China, India, and Malaysia. Asia is an endemic region, and the burden of talaromycosis is heavy, with prevalence ranging from 0.13 to 19,63% according to the different areas.

Ning et al. (2020) conducted a systemic review of talaromycosis cases published from 1964 to 2018. According to these authors, 288,000 accumulative cases and 87,900 deaths due to talaromycosis occurred globally at the end of 2018. From reports of 33 countries, the highest burden was in China (60.3%), followed by Thailand (30.4%) and Vietnam (8.4%). HIV-infected accounted for 89.9% of cases, 74.4% were male, and 0.5% were children.

Disseminated talaromycosis is more likely to develop in patients with HIV, resulting in higher mortality (80 to 100%) when not accompanied by accurate and prompt diagnosis and antifungal therapy. However, even when antifungal treatment is available, mortality can reach high rates, such as 30% (Li et al. 2022; Xu et al. 2023).

## *Scedosporiosis* and *Lomentosporiosis*

## *Etiology*

*Scedosporium apiospermum*, *Scedosporium aurantiacum*, and *Scedosporium boydii* (formerly *Pseudallescheria boydii*) are grouped in the *S. apiospermum* species complex. They are among the agents of mycetoma (when there is grain production at the site of infection) and scedosporiosis (formerly pseudallescheriasis—the other diseases caused by *Scedosporium* species) in humans.

*Lomentospora prolificans* (formerly *Scedosporium prolificans*) was reclassified to the genus *Lomenstospora* after genomic sequencing analyzes concluded that the fungus is not included in the genus *Scedosporium* (Konsoula et al. 2022b).

## *Ecology and Transmission*

These fungi are isolated from various environmental sources worldwide, such as soil, plants, fresh or polluted water, sewage, cattle and chicken manure, and other animals. However, the specific ecological niche for these fungi has not yet been fully recognized.

The form of disease acquisition is through inhalation of the pathogens into the lungs or paranal sinuses or after traumatic implantation into the skin of the fungus from the soil, water, surgery, and drug injections (Konsoula et al. 2022b).

## *Risk Factors*

*Lomentospora* and *Scedosporium* infections cause infections in immunocompetent and immunocompromised patients. Hosts with an intact immune system usually develop localized infections (e.g., osteoarticular, ocular, wounds, and onychomycosis) after inoculation of the fungus after trauma, including post-surgery (Hospenthal 2020b). In a review of 162 cases, Rodriguez-Tudela et al. (2009) described that 21% of patients did not have underlying diseases, but 82% were associated with surgery or trauma before acquiring the infection. These fungi were also identified as colonizing in patients with AIDS, cystic fibrosis, and cavitary lung disease (Lackner et al. 2011; Konsoula et al. 2022b; Hospenthal 2020b).

Immunocompromised patients are at high risk for invasive infections. Significant predisposing factors include organ transplantation (particularly hematopoietic stem cell transplant), hematologic malignancies (especially during periods of neutropenia), and cytoreductive chemotherapy. On the other hand, it is rarely observed in patients with HIV (Caira et al. 2008; Ramirez-Garcia et al. 2018; Konsoula et al. 2022b).

## *Incidence*

Among the fungal infections in organ transplant recipients, 5.66% of all invasive mycelial fungi were caused by *Lomentospora/Scedosporium* species (Husain et al. 2003). The high mortality rate among transplant recipients with lomentosporiosis/scedosporiosis can reach from around 50% to 87.3% (Caira et al. 2008; Husain et al. 2005; Konsoula et al. 2022a).

These diseases are associated with dry climates and have been reported in several countries, including Australia, the Southern USA, and European countries. Outbreaks have been described (Ramirez-Garcia et al. 2018; Idigoras et al. 2001; Lackner et al. 2011) (Delhaes et al. 2008). Konsoula et al. (2022a). reported that among 142 reported cases of lomentosporiosis, malignancies (72.5%), hemopoietic stem cell transplantation (23.2%), solid organ transplantation (16%), and AIDS (2%) were the primary underlying diseases observed. Neutropenia was present in 52% of patients.

## Eumycetoma Causative Agents

### *Etiology*

Mycetoma is a chronic granulomatous infection of cutaneous and subcutaneous tissues caused by bacteria (actinomycetoma) or fungi (eumycetoma). The disease typically affects the lower extremities, commonly a single foot, and is characterized by the triad of painless swelling, draining sinus with multiple discharging, and production of granules or grains (formed by the aggregations of the infectious organisms) (Musa et al. 2023). Several fungal species are agents of eumycetoma and can be classified into those that produce dark or black grains and those that form white to yellow (pale) grains (Table 2) (Hao et al. 2022; Nenoff et al. 2015; Hospenthal 2020a).

The most common agents of eumycetoma are *Madurella mycetomatis* (and the sibling species *M. pseudomycetomatis*, *M. fahalii*, and *M. tropicana*), *Trematosphaeria grisea* (formerly *Madurella grisea*), *Medicopsis romeroi* (formerly *Pyrenochaeta romeroi*), *Falciformispora senegalensis* (formerly *Leptosphaeria senegalensis*), *Scedosporium apiospermum* complex, and *Scedosporium boydii* complex (Emery and Denning 2020; Verma and Jha 2019; Hospenthal 2020a).

**Table 2** Typical eumycetoma grain colors

| Colors of grains | Infective agents |
|---|---|
| Dark and black grains | *Biatriospora* spp., *Corynespora cassicola*, *Curvularia geniculate*, *Curvularia lunata*, *Emarellia grisea*, *Emarellia paragrisea*, *Exophiala jeanselmei*, *Exophiala oligosperma*, *Falciformispora senegalensis* (formerly *Leptosphaeria senegalensis*), *Falciformispora thompkinsii*, *Madurella mycetomatis*, *Madurella fahalii*, *Madurella pseudomycetomatis*, *Madurella tropicana*, *Medicopsis* (formerly *Pyrenochaeta*) *mackinnonii*, *Medicopsis romeroi* (formerly *P. Romeroi*), *Phaeoacremonium* spp., *Phialophora verrucosa*, *Plenodomas avramii*, *Pseudochaetosphaeroma* spp., *Rhinocladiella atrovirens*, *Rhytidhysteron* spp., *Roussoella* spp., *Trematosphaeria grisea* (formerly *Madurella grisea*) |
| White to yellow (pale grains) | *Acremonium recifei*, *aspergillus flavus*, *aspergillus hollandicus*, *aspergillus nidulans*, *Cladosporium cladosporioides*, *Cylindrocarpon destructans*, *Diaporthe phaseolorum*, *Fusarium (formerly Acremonium) falciforme*, *Fusarium keratoplasticum*, *Fusarium moniliforme*, *Fusarium oxysoprum*, *Fusarium pseudensiforme*, *Fusarium solani*, *Microsporum audouinii*, *Microsporum ferrugineum*, *Microsporum langeronii*, *Neotestudina rosatii*, *Phialophora* (formerly *Cylindrocarpon*) *cyanescens*, *Pleurostomophora ochracea*, *Polycytella hominis*, *Sarocladium* (formerly *Acremonium*), *Scedosporium apiospermum* complex, and *Scedosporium boydii* complex |

Data was compiled from Nenoff et al. (2015), Hospenthal (2020a), and (Hao et al. 2022)

## Ecology and Transmission

The microorganisms that are agents of eumycetoma have been found in plants, soil, and water (tropical and subtropical climates) in the endemic regions. The infection is believed to result from the implantation of the microorganism in the cutaneous and subcutaneous tissue after minor trauma. After local tissue incubation, the microorganism proliferates, and the disease progress to visible lesions (Hao et al. 2022; Sow et al. 2020; Musa et al. 2023).

## Risk Factors

Specific risk factors for the development of eumycetoma are unknown. However, some aspects of the environment and lifestyle are likely involved. The main predisposing factors are related to the geographic location (eumycetoma is more common in tropical and subtropical regions, particularly in Africa, Asia, and Latin America, in an area known as the "mycetoma belt") (Fahal 2004; Emery and Denning 2020), occupational exposure (several patients with eumycetoma work in agricultural jobs and construction, which may increase the risk of exposition to microorganisms in the environment) (Emery and Denning 2020), trauma (injuries in cutaneous and subcutaneous tissue can increase the risk of eumycetoma, as they provide an entry point for the fungi) (Nenoff et al. 2015), immunodeficiency and genetic susceptibility (individuals with a weakened immune system are at higher risk of developing the more aggressive disease. Genetic factors may play a role in the development of eumycetoma) (Ali et al. 2020).

## Incidence

The exact global burden of mycetoma is not concisely known. Fungal species' prevalence depends on the geographic region, climate, humidity, and socioeconomic and health education status. Highest incidences are described in remote rural areas from tropical and subtropical countries in the "mycetoma belt" regions (altitudes between 30° North and 15° South), particularly in Sudan, Somalia, Senegal, Yemen, India, Venezuela, and Mexico (Hao et al. 2022; Musa et al. 2023).

Mucormycosis is uncommon in children. However, in Sudan, 722 cases were reported between 1991 and 2009. Age ranged from 4 to 17 years, with the majority (75.5%) being male (Fahal and Sabaa 2010). Although it can reach all ages, a higher incidence is reported in adults aged between 20 and 40 years and who have activities related to the environment (Zijlstra et al. 2016).

In a systematic review and meta-analysis, van de Sande (2013) evaluated 8763 cases of mycetoma. Most reports were from Mexico, Sundan, and India. Prevalence of 1.81 and 3.49 cases per 100,000 inhabitants was related in Mauritania and Sudan, respectively. Most cases (70%) occurred in patients aged 11 to 40 and men (77%). *M. mycetomatis* was the most prevalent causative agent worldwide.

More recently, Emery and Denning (2020) described the global distribution of actinomycetoma and eumycetoma after a literature review of 19,494 cases dating from 1876 to 2019. Most patients were also from Mexico, India, and Sudan. However, the authors identified new and emerging geographical loci, including the United States, Venezuela, Italy, Australia, and China. More eumycetoma cases were reported in some central African countries.

## Conclusion

Fungal infections pose a serious public health threat, especially to vulnerable populations with weakened immune systems. The rising incidence of fungal pathogens, including the emergence of multidrug-resistant species like *C. auris*, highlights the urgent need for improved diagnostics and more effective antifungal therapies. Epidemiological studies show that these infections are responsible for significant levels of illness and death, with patterns that vary depending on geography and species distribution. Environmental conditions, healthcare practices, and individual patient risks complicate managing fungal diseases.

Identifying high-risk populations, such as organ transplant recipients, cancer patients, or individuals living with HIV/AIDS, is critical to early intervention and the development of targeted prevention strategies. Additionally, the rising number of opportunistic fungal infections connected to global health crises, like the COVID-19 pandemic, underscores the importance of continuous monitoring and research to understand how these infections spread and can be controlled.

Effectively addressing the burden of fungal infections necessitates a comprehensive approach. This approach should integrate epidemiological insights, heightened clinical awareness, and the latest advances in antifungal treatment. By adopting such a strategy, we can more effectively confront the growing threat posed by these intricate and often fatal diseases.

## References

Adenis AA, Valdes A, Cropet C, Mccotter OZ, Derado G, Couppie P, Chiller T, Nacher M (2018) Burden of HIV-associated histoplasmosis compared with tuberculosis in Latin America: a modelling study. *Lancet Infect Dis* 18:1150–1159

Alastruey-Izquierdo A, Mellado E, Cuenca-Estrella M (2012) Current section and species complex concepts in *Aspergillus*: recommendations for routine daily practice. *Ann N Y Acad Sci* 1273:18–24

Al-Hatmi AM, Hagen F, Menken SB, Meis JF, de Hoog GS (2016a) Global molecular epidemiology and genetic diversity of *Fusarium*, a significant emerging group of human opportunists from 1958 to 2015. *Emerg Microbes Infect* 5:e124

Al-Hatmi AM, Meis JF, de Hoog GS (2016b) *Fusarium*: molecular diversity and intrinsic drug resistance. *PLoS Pathog* 12:e1005464

Ali RS, Newport MJ, Bakhiet SM, Ibrahim ME, Fahal AH (2020) Host genetic susceptibility to mycetoma. *PLoS Negl Trop Dis* 14:e0008053

Ames R, Brown AJP, Gudelj I, Nev OA (2023) Analysis of *Pneumocystis* transcription factor evolution and implications for biology and lifestyle. *mBio*:e0271122

Ampel NM (2020) Coccidioidomycosis: changing concepts and knowledge gaps. *J Fungi (Basel)* 6:354

Andresen D, Donaldson A, Choo L, Knox A, Klaassen M, Ursic C, Vonthethoff L, Krilis S, Konecny P (2005) Multifocal cutaneous mucormycosis complicating polymicrobial wound infections in a tsunami survivor from Sri Lanka. *Lancet* 365:876–878

Apostolopoulou A, Fishman JA (2022) The pathogenesis and diagnosis of *Pneumocystis jiroveci* pneumonia. *J Fungi (Basel)* 8:1167

Arauz AB, Papineni P (2021) Histoplasmosis. *Infect Dis Clin North Am* 35:471–491

Armstrong-James D, Meintjes G, Brown GD (2014) A neglected epidemic: fungal infections in HIV/AIDS. *Trends Microbiol* 22:120–127

Azie N, Neofytos D, Pfaller M, Meier-Kriesche H-U, Quan S-P, Horn D (2012) The PATH (Prospective Antifungal Therapy) Alliance® registry and invasive fungal infections: update 2012. Diagn Microbiol Infect Dis 73:293–300

Balajee SA, Houbraken J, Verweij PE, Hong SB, Yaghuchi T, Varga J, Samson RA (2007) *Aspergillus* species identification in the clinical setting. *Stud Mycol* 59:39–46

Barros N, Wheat JL, Hage C (2023) Pulmonary histoplasmosis: a clinical update. J Fungi 9

Batista BG, Chaves MA, Reginatto P, Saraiva OJ, Fuentefria AM (2020) Human fusariosis: an emerging infection that is difficult to treat. *Rev Soc Bras Med Trop* 53:e20200013

Bellissimo-Rodrigues F, Machado AA, Martinez R (2011) Paracoccidioidomycosis epidemiological features of a 1,000-cases series from a hyperendemic area on the southeast of Brazil. *Am J Trop Med Hyg* 85:546–550

Bellmann-Weiler R, Bellmann R (2019) Clinical syndromes: *Aspergillus*. In: Presterl E (ed) Clinically relevant mycoses, 1st edn. Springer, New York, NY

Benard G (2021) Pathogenesis and classification of paracocidioidomycosis: new insights from old good stuff. *Open Forum Infect Dis* 8:ofaa624

Benedict K, Mody RK (2016) Epidemiology of histoplasmosis outbreaks, United States, 1938–2013. *Emerg Infect Dis* 22:370–378

Benedict K, Thompson GR 3rd, Deresinski S, Chiller T (2015) Mycotic infections acquired outside areas of known endemicity, United States. *Emerg Infect Dis* 21:1935–1941

Bitar D, Van Cauteren D, Lanternier F, Dannaoui E, Che D, Dromer F, Desenclos JC, Lortholary O (2009) Increasing incidence of zygomycosis (mucormycosis), France, 1997–2006. *Emerg Infect Dis* 15:1395–1401

Bongomin F, Gago S, Oladele RO, Denning DW (2017) Global and multi-national prevalence of fungal diseases-estimate precision. *J Fungi (Basel)* 3:57

Cadena J, Thompson GR 3rd, Patterson TF (2021) Aspergillosis: epidemiology, diagnosis, and treatment. *Infect Dis Clin North Am* 35:415–434

Caira M, Girmenia C, Valentini CG, Sanguinetti M, Bonini A, Rossi G, Fianchi L, Leone G, Pagano L (2008) Scedosporiosis in patients with acute leukemia: a retrospective multicenter report. *Haematologica* 93:104–110

Cao C, Xi L, Chaturvedi V (2019) Talaromycosis (Penicilliosis) due to *Talaromyces* (*Penicillium*) *marneffei*: insights into the clinical trends of a major fungal disease 60 years after the discovery of the pathogen. *Mycopathologia* 184:709–720

Casadevall A (2018) Fungal diseases in the 21st century: the near and far horizons. *Pathog Immun* 3:183–196

Castro-Fuentes CA, Reyes-Montes MDR, Frias-de-Leon MG, Valencia-Ledezma OE, Acosta-Altamirano G, Duarte-Escalante E (2022) *Aspergillus*-SARS-CoV-2 coinfection: what is known? *Pathogens* 11:1227

CDC (2023) *Valley fever (coccidioidomycosis) statistics* [Online]. Centers for Disease Control and Prevention. Available: https://www.cdc.gov/fungal/diseases/coccidioidomycosis/statistics.html. Accessed 19 Feb 2023

Chakrabarti A, Kaur H, Savio J, Rudramurthy SM, Patel A, Shastri P, Pamidimukkala U, Karthik R, Bhattacharya S, Kindo AJ, Bhattacharya P, Todi S, Gopalakrishan R, Singh P, Pandey A, Agarwal R (2019) Epidemiology and clinical outcomes of invasive mould infections in Indian intensive care units (FISF study). *J Crit Care* 51:64–70

Challa S (2018) Pathogenesis and pathology of invasive aspergillosis. *Curr Fungal Infect Rep* 12:23–32

Challa, S. 2019. Mucormycosis: pathogenesis and pathology. *Curr Fungal Infect Rep 13 11-20*

Chan JF, Chan TS, Gill H, Lam FY, Trendell-Smith NJ, Sridhar S, Tse H, Lau SK, Hung IF, Yuen KY, Woo PC (2015) Disseminated infections with *Talaromyces marneffei* in non-AIDS patients given monoclonal antibodies against CD20 and kinase Inhibitors. *Emerg Infect Dis* 21:1101–1106

Chen L, Han X, Li Y, Zhang C, Xing X (2020) Invasive pulmonary aspergillosis in immunocompetent patients hospitalised with influenza A-related pneumonia: a multicenter retrospective study. *BMC Pulm Med* 20:239

Chuleerarux N, Thongkam A, Manothummetha K, Nematollahi S, Dioverti-Prono V, Torvorapanit P, Langsiri N, Worasilchai N, Plongla R, Chindamporn A, Sanguankeo A, Permpalung N (2021) Does post-transplant cytomegalovirus increase the risk of invasive aspergillosis in solid organ transplant recipients? A systematic review and meta-analysis. *J Fungi (Basel)* 7:327

Cisse OH, Hauser PM (2018) Genomics and evolution of *Pneumocystis* species. *Infect Genet Evol* 65:308–320

Cleveland AA, Farley MM, Harrison LH, Stein B, Hollick R, Lockhart SR, Magill SS, Derado G, Park BJ, Chiller TM (2012) Changes in incidence and antifungal drug resistance in candidemia: results from population-based laboratory surveillance in Atlanta and Baltimore, 2008–2011. *Clin Infect Dis* 55:1352–1361

Colombo AL, Tobon A, Restrepo A, Queiroz-Telles F, Nucci M (2011) Epidemiology of endemic systemic fungal infections in Latin America. *Med Mycol* 49:785–798

Coutinho ZF, Silva D, Lazera M, Petri V, Oliveira RM, Sabroza PC, Wanke B (2002) Paracoccidioidomycosis mortality in Brazil (1980–1995). *Cad Saude Publica* 18:1441–1454

Coutinho ZF, Wanke B, Travassos C, Oliveira RM, Xavier DR, Coimbra CE Jr (2015) Hospital morbidity due to paracoccidioidomycosis in Brazil (1998–2006). *Trop Med Int Health* 20:673–680

Cushion MT, Tisdale-Macioce N, Sayson SG, Porollo A (2021) The persistent challenge of *Pneumocystis* growth outside the mammalian lung: past and future approaches. *Front Microbiol* 12:681474

Dadar M, Tiwari R, Karthik K, Chakraborty S, Shahali Y, Dhama K (2018) *Candida albicans* – biology, molecular characterization, pathogenicity, and advances in diagnosis and control – an update. *Microb Pathog* 117:128–138

Dagenais TR, Keller NP (2009) Pathogenesis of *Aspergillus fumigatus* in invasive aspergillosis. *Clin Microbiol Rev* 22:447–465

Davis MR, Thompson GR 3rd, Patterson TF (2020) Fungal infections potentiated by biologics. *Infect Dis Clin North Am* 34:389–411

de Macedo PM, Scramignon-Costa BS, Almeida-Paes R, Trilles L, de Oliveira LSC, Zancope-Oliveira RM, do Valle ACF, Wanke B (2020) *Paracoccidioides brasiliensis* habitat: far beyond armadillo burrows? *Mem Inst Oswaldo Cruz* 115:e200208

de Perio MA, Materna BL, Sondermeyer Cooksey GL, Vugia DJ, Su CP, Luckhaupt SE, Mcnary J, Wilken JA (2019) Occupational coccidioidomycosis surveillance and recent outbreaks in California. *Med Mycol* 57:S41–s45

Delhaes L, Harun A, Chen SC, Nguyen Q, Slavin M, Heath CH, Maszewska K, Halliday C, Robert V, Sorrell TC, Australian scedosporium study Group, Meyer W (2008) Molecular typing of Australian *Scedosporium* isolates showing genetic variability and numerous *S. aurantiacum*. *Emerg Infect Dis* 14:282–290

Desoubeaux G, Coste AT, Imbert C, Hennequin C (2022) Overview about *Candida auris*: what's up 12 years after its first description? *J Mycol Med* 32:101248

Diaz JH (2018) Environmental and wilderness-related risk factors for histoplasmosis: more than bats in caves. *Wilderness Environ Med* 29:531–540

Didehdar M, Chegini Z, Moradabadi A, Anoushirvani AA, Tabaeian SP, Yousefimashouf M, Shariati A (2022) Gastrointestinal mucormycosis: a periodic systematic review of case reports from 2015 to 2021. *Microb Pathog* 163:105388

Dignani MC (2014) Epidemiology of invasive fungal diseases on the basis of autopsy reports. *F1000Prime Rep* 6:81

Diniz-Lima I, Fonseca LMD, Silva-Junior EBD, Guimarães-DE-Oliveira JC, Freire-de-Lima L, Nascimento DO, Morrot A, Previato JO, Mendonça-Previato L, Decote-Ricardo D, Freire-de-Lima CG (2022) *Cryptococcus*: history, epidemiology and immune evasion. *Applied Sciences* 12:7086

Dohn MN, White ML, Vigdorth EM, Ralph Buncher C, Hertzberg VS, Baughman RP, George Smulian A, Walzer PD (2000) Geographic clustering of *Pneumocystis carinii* pneumonia in patients with HIV infection. *Am J Respir Crit Care Med* 162:1617–1621

Emery D, Denning DW (2020) The global distribution of actinomycetoma and eumycetoma. *PLoS Negl Trop Dis* 14:e0008397

Fahal AH (2004) Mycetoma: a thorn in the flesh. *Trans R Soc Trop Med Hyg* 98:3–11

Fahal AH, Sabaa AH (2010) Mycetoma in children in Sudan. *Trans R Soc Trop Med Hyg* 104:117–121

Falci DR, Monteiro AA, Braz Caurio CF, Magalhães TCO, Xavier MO, Basso RP, Melo M, Schwarzbold AV, Ferreira PRA, Vidal JE, Marochi JP, Godoy CSM, Soares RBA, Paste A, Bay MB, Pereira-Chiccola VL, Damasceno LS, Leitão T, Pasqualotto AC (2019) Histoplasmosis, an underdiagnosed disease affecting people living with HIV/AIDS in Brazil: results of a multicenter prospective cohort study using both classical mycology tests and histoplasma urine antigen detection. *Open Forum Infect Dis* 6:ofz073

Falcone M, Massetti AP, Russo A, Vullo V, Venditti M (2011) Invasive aspergillosis in patients with liver disease. *Med Mycol* 49:406–413

Fernandez-Pittol M, Alejo-Cancho I, Rubio-Garcia E, Cardozo C, Puerta-Alcalde P, Moreno-Garcia E, Garcia-Pouton N, Garrido M, Villanueva M, Alastruey-Izquierdo A, Pitart C, Garcia-Vidal C, Marco F (2022) Aspergillosis by cryptic *Aspergillus* species: a case series and review of the literature. *Rev Iberoam Micol* 39:44–49

Fisher FS, Bultman MW, Johnson SM, Pappagianis D, Zaborsky E (2007) *Coccidioides* niches and habitat parameters in the southwestern United States: Q matter of scale. *Ann N Y Acad Sci* 1111:47–72

Fishman, J. A. 2020. *Pneumocystis jiroveci*. Semin Respir Crit Care Med, 41, 141–157

Forsberg K, Woodworth K, Walters M, Berkow EL, Jackson B, Chiller T, Vallabhaneni S (2019) *Candida auris*: the recent emergence of a multidrug-resistant fungal pathogen. *Med Mycol* 57:1–12

Galgiani JN (2020) Coccidioidomycosis (*Coccidioides* species). In: Bennett JE, Dolin R, Blaser MJ (eds) Mandell, Douglas, and Bennett's principles and practice of infectious diseases, 9th edn. Elsevier, Philadelphia

Garvey M, Rowan NJ (2023) Pathogenic drug resistant fungi: a review of mitigation strategies. *Int J Mol Sci* 24

Geddes-Mcalister J, Shapiro RS (2019) New pathogens, new tricks: emerging, drug-resistant fungal pathogens and future prospects for antifungal therapeutics. *Ann N Y Acad Sci* 1435:57–78

Geremia N, Brugnaro P, Solinas M, Scarparo C, Panese S (2023) *Candida auris* as an emergent public health problem: a current update on European outbreaks and cases. *Healthcare (Basel)* 11:425

Goldberg DW, Tenforde MW, Mitchell HK, Jarvis JN (2018) Neurological sequelae of adult meningitis in Africa: a systematic literature review. *Open Forum Infect Dis* 5:ofx246

Guarro J (2013) Fusariosis, a complex infection caused by a high diversity of fungal species refractory to treatment. *Eur J Clin Microbiol Infect Dis* 32:1491–1500

Guery R, Lanternier F, Lortholary O (2019) Clinical Syndromes: cryptococcosis. In: Presterl E (ed) *Clinically relevant mycoses*, 1st edn. Springer, New York, NY

Guinea J, Escribano P, Vena A, Munoz P, Martinez-Jimenez MDC, Padilla B, Bouza E (2017) Increasing incidence of mucormycosis in a large Spanish hospital from 2007 to 2015: epidemiology and microbiological characterization of the isolates. *PLoS One* 12:e0179136

Gunathilaka M, Senevirathna R, Illappereruma SC, Keragala K, Hathagoda KLW, Bandara H (2023) Mucormycosis-causing fungi in humans: a meta-analysis establishing the phylogenetic relationships using internal transcribed spacer (ITS) sequences. *J Med Microbiol* 72:001651

Gupta P, Kaur H, Kenwar DB, Gupta P, Agnihotri S, Rudramurthy SM (2022) First case of subcutaneous infection by *Talaromyces marneffei* in a renal transplant recipient from India and review of literature. *J Mycol Med* 32:101207

Hagen F, Khayhan K, Theelen B, Kolecka A, Polacheck I, Sionov E, Falk R, Parnmen S, Lumbsch HT, Boekhout T (2015) Recognition of seven species in the *Cryptococcus gattii/Cryptococcus neoformans* species complex. *Fungal Genet Biol* 78:16–48

Hahn RC, Hagen F, Mendes RP, Burger E, Nery AF, Siqueira NP, Guevara A, Rodrigues AM, Camargo DE, Z. P. (2022) Paracoccidioidomycosis: current status and future trends. *Clin Microbiol Rev* 35:e0023321

Hajjeh RA, Conn LA, Stephens DS, Baughman W, Hamill R, Graviss E, Pappas PG, Thomas C, Reingold A, Rothrock G, Hutwagner LC, Schuchat A, Brandt ME, Pinner RW (1999) Cryptococcosis: population-based multistate active surveillance and risk factors in human immunodeficiency virus-infected persons. Cryptococcal Active Surveillance Group. *J Infect Dis* 179:449–454

Hao X, Cognetti M, Burch-Smith R, Mejia EO, Mirkin G (2022) Mycetoma: development of diagnosis and treatment. *J Fungi (Basel)* 8:743

Hasmoni MH, Shah AS, Ayoub S, Hin LS, Abd Rashid MA (2010) Disseminated histoplasmosis presenting with chronic ulcerative tongue lesions in a patient with diabetes. *BMJ Case Rep*:2010

Hassan MIA, Voigt K (2019) Pathogenicity patterns of mucormycosis: epidemiology, interaction with immune cells and virulence factors. *Med Mycol* 57:S245–S256

Hauser PM (2021) *Pneumocystis* mating-type locus and sexual cycle during infection. *Microbiol Mol Biol Rev* 85:e0000921

Hohmann FB, Chaves RCF, Olivato GB, Souza GM, Galindo VB, Silva M Jr, Martino MDV, Menezes FG, Correa TD (2023) Characteristics, risk factors, and outcomes of bloodstream *Candida* infections in the intensive care unit: a retrospective cohort study. *J Int Med Res* 51:3000605221131122

Hospenthal DR (2020a) Agents of mycetoma. In: Bennett JE, Dolin R, Blaser MJ (eds) *Mandell, douglas, and bennett's principles and practice of infectious diseases*, 9th edn. Elsevier, Philadelphia

Hospenthal DR (2020b) Uncommon fungi and related species. In: Bennett JE, Dolin R, Blaser MJ (eds) *Mandell, douglas, and bennett's principles and practice of infectious diseases*, 9th edn. Elsevier, Philadelphia

Husain S, Alexander BD, Munoz P, Avery RK, Houston S, Pruett T, Jacobs R, Dominguez EA, Tollemar JG, Baumgarten K, Yu CM, Wagener MM, Linden P, Kusne S, Singh N (2003) Opportunistic mycelial fungal infections in organ transplant recipients: emerging importance of non-*Aspergillus* mycelial fungi. *Clin Infect Dis* 37:221–229

Husain S, Munoz P, Forrest G, Alexander BD, Somani J, Brennan K, Wagener MM, Singh N (2005) Infections due to *Scedosporium apiospermum* and *Scedosporium prolificans* in transplant recipients: clinical characteristics and impact of antifungal agent therapy on outcome. *Clin Infect Dis* 40:89–99

Idigoras P, Perez-Trallero E, Pineiro L, Larruskain J, Lopez-Lopategui MC, Rodriguez N, Gonzalez JM (2001) Disseminated infection and colonization by *Scedosporium prolificans*: a review of 18 cases, 1990–1999. *Clin Infect Dis* 32:E158–E165

Imbert S, Cassaing S, Bonnal C, Normand AC, Gabriel F, Costa D, Blaize M, Lachaud L, Hasseine L, Kristensen L, Guitard J, Schuttler C, Raberin H, Brun S, Hendrickx M, Piarroux R, Fekkar A (2021) Invasive aspergillosis due to *Aspergillus* cryptic species: a prospective multicentre study. *Mycoses* 64:1346–1353

Jeong W, Keighley C, Wolfe R, Lee WL, Slavin MA, Kong DCM, Chen SC (2019) The epidemiology and clinical manifestations of mucormycosis: a systematic review and meta-analysis of case reports. *Clin Microbiol Infect* 25:26–34

Kandemir H, Dukik K, de Melo Teixeira M, Stielow JB, Delma FZ, AL-Hatmi AMS, Ahmed SA, Ilkit M, De Hoog GS (2022) Phylogenetic and ecological reevaluation of the order Onygenales. *Fungal Divers* 115:1–72

Kauffman CA (2008) Diagnosis of histoplasmosis in immunosuppressed patients. *Curr Opin Infect Dis* 21:421–425

Khawcharoenporn T, Apisarnthanarak A, Mundy LM (2007) Non-*neoformans* cryptococcal infections: a systematic review. *Infection* 35:51–58

Kidd SE, Abdolrasouli A, Hagen F (2023) Fungal nomenclature: managing change is the name of the game. *Open Forum Infect Dis* 10:ofac559

Kirkland TN, Stevens DA, Hung CY, Beyhan S, Taylor JW, Shubitz LF, Duttke SH, Heidari A, Johnson RH, Deresinski SC, Lauer A, Fierer J (2022) Coccidioides species: a review of basic research: 2022. *J Fungi (Basel)* 8:859

Kluge S, Strauss R, Kochanek M, Weigand MA, Rohde H, Lahmer T (2021) Aspergillosis: emerging risk groups in critically ill patients. *Med Mycol* 60:myab064

Konsoula A, Agouridis AP, Markaki L, Tsioutis C, Spernovasilis N (2022a) *Lomentospora prolificans* disseminated infections: a systematic review of reported cases. *Pathogens* 12

Konsoula A, Tsioutis C, Markaki I, Papadakis M, Agouridis AP, Spernovasilis N (2022b) *Lomentospora prolificans*: an emerging opportunistic fungal pathogen. *Microorganisms* 10

Kontoyiannis DP, Bodey GP (2002) Invasive aspergillosis in 2002: an update. *Eur J Clin Microbiol Infect Dis* 21:161–172

Kontoyiannis DP, Marr KA, Park BJ, Alexander BD, Anaissie EJ, Walsh TJ, Ito J, Andes DR, Baddley JW, Brown JM, Brumble LM, Freifeld AG, Hadley S, Herwaldt LA, Kauffman CA, Knapp K, Lyon GM, Morrison VA, Papanicolaou G, Patterson TF, Perl TM, Schuster MG, Walker R, Wannemuehler KA, Wingard JR, Chiller TM, Pappas PG (2010) Prospective surveillance for invasive fungal infections in hematopoietic stem cell transplant recipients, 2001–2006: overview of the transplant-associated infection surveillance network (TRANSNET) database. *Clin Infect Dis* 50:1091–1100

Kontoyiannis DP, Yang H, Song J, Kelkar SS, Yang X, Azie N, Harrington R, Fan A, Lee E, Spalding JR (2016) Prevalence, clinical and economic burden of mucormycosis-related hospitalizations in the United States: a retrospective study. *BMC Infect Dis* 16:730

Krajicek BJ, Thomas CF Jr, Limper AH (2009) *Pneumocystis* pneumonia: current concepts in pathogenesis, diagnosis, and treatment. *Clin Chest Med* 30(265–78):vi

Kullberg BJ, Arendrup MC (2015) Invasive candidiasis. *N Engl J Med* 373:1445–1456

Kumar S, Kumar A, Roudbary M, Mohammadi R, Cernakova L, Rodrigues CF (2022) Overview on the infections related to rare *Candida* species. *Pathogens* 11

Kwon-Chung KJ, Bennett JE, Wickes BL, Meyer W, Cuomo CA, Wollenburg KR, Bicanic TA, Castaneda E, Chang YC, Chen J, Cogliati M, Dromer F, Ellis D, Filler SG, Fisher MC, Harrison TS, Holland SM, Kohno S, Kronstad JW, Lazera M, Levitz SM, Lionakis MS, May RC, Ngamskulrongroj P, Pappas PG, Perfect JR, Rickerts V, Sorrell TC, Walsh TJ, Williamson PR,

Xu J, Zelazny AM, Casadevall A (2017) The case for adopting the "species complex" nomenclature for the etiologic agents of cryptococcosis. *mSphere* 2:e00357–e00316
Lackner M, Rezusta A, Villuendas MC, Palacian MP, Meis JF, Klaassen CH (2011) Infection and colonisation due to *Scedosporium* in Northern Spain. An *in vitro* antifungal susceptibility and molecular epidemiology study of 60 isolates. *Mycoses* 54(Suppl 3):12–21
Lanternier F, Dannaoui E, Morizot G, Elie C, Garcia-Hermoso D, Huerre M, Bitar D, Dromer F, Lortholary O, French Mycosis Study Group (2012) A global analysis of mucormycosis in France: the RetroZygo Study (2005–2007). *Clin Infect Dis* 54(Suppl 1):S35–S43
Latge JP, Chamilos G (2019) *Aspergillus fumigatus* and aspergillosis in 2019. *Clin Microbiol Rev* 33:e00140–e00118
Levesque E, Ait-Ammar N, Dudau D, Clavieras N, Feray C, Foulet F, Botterel F (2019) Invasive pulmonary aspergillosis in cirrhotic patients: analysis of a 10-year clinical experience. *Ann Intensive Care* 9:31
Li Z, Denning DW (2023) The impact of corticosteroids on the outcome of fungal disease: a systematic review and meta-analysis. *Curr Fungal Infect Rep*:1–17
Li D, Liang H, Zhu Y, Chang Q, Pan P, Zhang Y (2022) Clinical characteristics, laboratory findings, and prognosis in patients with *Talaromyces marneffei* infection across various immune statuses. *Front Med (Lausanne)* 9:841674
Limper AH, Adenis A, Le T, Harrison TS (2017) Fungal infections in HIV/AIDS. *Lancet Infect Dis* 17:e334–e343
Linder KA, Kauffman CA (2019) Histoplasmosis: epidemiology, diagnosis, and clinical manifestations. *Curr Fungal Infect Rep* 13:120–128
Ling F, Guo T, Li J, Chen Y, Xu M, Li S, Zhu L (2022) Gastrointestinal *Talaromyces marneffei* infection in a patient with AIDS: a case report and systematic review. *Front Immunol* 13:980242
Lionakis MS, Edwards JR, J. E. (2020) *Candida* species. In: Bennett JE, Dolin R, Blaser MJ (eds) *Mandell, douglas, and bennett's principles and practice of infectious diseases*, 9th edn. Elsevier, Philadelphia
Lockhart SR, Etienne KA, Vallabhaneni S, Farooqi J, Chowdhary A, Govender NP, Colombo AL, Calvo B, Cuomo CA, Desjardins CA, Berkow EL, Castanheira M, Magobo RE, Jabeen K, Asghar RJ, Meis JF, Jackson B, Chiller T, Litvintseva AP (2017) Simultaneous emergence of multidrug-resistant *Candida auris* on 3 continents confirmed by whole-genome sequencing and epidemiological analyses. *Clin Infect Dis* 64:134–140
Ma L, Cisse OH, Kovacs JA (2018) A molecular window into the biology and epidemiology of *Pneumocystis* spp. *Clin Microbiol Rev* 31:e00009–e00018
Magill SS, Edwards JR, Bamberg W, Beldavs ZG, Dumyati G, Kainer MA, Lynfield R, Maloney M, Mcallister-Hollod L, Nadle J, Ray SM, Thompson DL, Wilson LE, Fridkin SK, Emerging infections program healthcare-associated infections & Antimicrobial use prevalence survey team (2014) Multistate point-prevalence survey of health care-associated infections. *N Engl J Med* 370:1198–1208
Marr KA, Sun Y, Spec A, Lu N, Panackal A, Bennett J, Pappas P, Ostrander D, Datta K, Zhang SX, Williamson PR, Cryptococcus Infection Network Cohort Study Working Group (2019) A multicenter, longitudinal cohort study of cryptococcosis in HIV-negative people in the United States. *Clin Infect Dis*
Martinez R (2015) Epidemiology of paracoccidioidomycosis. *Rev Inst Med Trop Sao Paulo* 57(Suppl 19):11–20
Martinez R (2017) New trends in paracoccidioidomycosis epidemiology. *J Fungi (Basel)* 3:1
May RC, Stone NR, Wiesner DL, Bicanic T, Nielsen K (2016) *Cryptococcus*: from environmental saprophyte to global pathogen. *Nat Rev Microbiol* 14:106–117
Maziarz EK, Perfect JR (2016) Cryptococcosis. *Infect Dis Clin North Am* 30:179–206
Mcneil JI, Kan VL (1995) Decline in the incidence of cryptococcosis among HIV-infected patients. *J Acquir Immune Defic Syndr Hum Retrovirol* 9:206–208

Medrano FJ, Respaldiza N, Medrano A, Varela JM, Montes-Cano M, de la Horra C, Wichmann I, Ferrer S, Calderon E (2003) Seroprevalence of *Pneumocystis* human infection in southern Spain. *J Eukaryot Microbiol* 50(Suppl):649–650

Medrano FJ, Montes-Cano M, Conde M, de la Horra C, Respaldiza N, Gasch A, Perez-Lozano MJ, Varela JM, Calderon EJ (2005) *Pneumocystis jirovecii* in general population. *Emerg Infect Dis* 11:245–250

Mendes RP, Cavalcante RS, Marques SA, Marques MEA, Venturini J, Sylvestre TF, Paniago AMM, Pereira AC, da Silva JF, Fabro AT, Bosco SMG, Bagagli E, Hahn RC, Levorato AD (2017) Paracoccidioidomycosis: current perspectives from Brazil. *Open Microbiol J* 11:224–282

Miller RF, Evans HE, Copas AJ, Cassell JA (2007) Climate and genotypes of *Pneumocystis jirovecii*. *Clin Microbiol Infect* 13:445–448

Mina S, Yaakoub H, Annweiler C, Dubee V, Papon N (2022) COVID-19 and Fungal infections: a double debacle. *Microbes Infect* 24:105039

Montes-Cano MA, Chabe M, Fontillon-Alberdi M, De-Lahorra C, Respaldiza N, Medrano FJ, Varela JM, Dei-Cas E, Calderon EJ (2009) Vertical transmission of *Pneumocystis jirovecii* in humans. *Emerg Infect Dis* 15:125–127

Musa EA, Abdoon IH, Bakhiet SM, Osman B, Abdalla SA, Fahal AH (2023) Mycetoma management and clinical outcomes: the Mycetoma Research Center experience. *Trans R Soc Trop Med Hyg* 117:12–21

Nam H, Ison MG (2020) 1496. Aspergillosis complicating severe respiratory syncytial virus (RSV) in ICU patients: a retrospective cohort study. *Open Forum Infect Dis* 7:S750

Narayanasamy S, Dat VQ, Thanh NT, Ly VT, Chan JF, Yuen KY, Ning C, Liang H, Li L, Chowdhary A, Youngchim S, Supparatpinyo K, Aung NM, Hanson J, Andrianopoulos A, Dougherty J, Govender NP, Denning DW, Chiller T, Thwaites G, van Doorn HR, Perfect J, Le T (2021) A global call for talaromycosis to be recognised as a neglected tropical disease. *Lancet Glob Health* 9:e1618–e1622

Neblett Fanfair R, Benedict K, Bos J, Bennett SD, Lo YC, Adebanjo T, Etienne K, Deak E, Derado G, Shieh WJ, Drew C, Zaki S, Sugerman D, Gade L, Thompson EH, Sutton DA, Engelthaler DM, Schupp JM, Brandt ME, Harris JR, Lockhart SR, Turabelidze G, Park BJ (2012) Necrotizing cutaneous mucormycosis after a tornado in Joplin, Missouri, in 2011. *N Engl J Med* 367:2214–2225

Nenoff, P., van de Sande, W. W., Fahal, A. H., Reinel, D. & Schofer, H. 2015. Eumycetoma and actinomycetoma--an update on causative agents, epidemiology, pathogenesis, diagnostics and therapy. *J Eur Acad Dermatol Venereol,* 29**,** 1873–1883

Ning C, Wei W, Xu B (2020) The global distribution, drivers, and burden of talaromycosis 1964–2018. In: Conference of Retrovirus and Opportunistic Infections. DukeHealth, 8–11

Noble JA, Nelson RG, Fufaa GD, Kang P, Shafir SC, Galgiani JN (2016) Effect of geography on the analysis of coccidioidomycosis-associated deaths, United States. *Emerg Infect Dis* 22:1821–1823

Nucci M, Marr KA, Vehreschild MJ, de Souza CA, Velasco E, Cappellano P, Carlesse F, Queiroz-Telles F, Sheppard DC, Kindo A, Cesaro S, Hamerschlak N, Solza C, Heinz WJ, Schaller M, Atalla A, Arikan-Akdagli S, Bertz H, Galvao Castro C Jr, Herbrecht R, Hoenigl M, Harter G, Hermansen NE, Josting A, Pagano L, Salles MJ, Mossad SB, Ogunc D, Pasqualotto AC, Araujo V, Troke PF, Lortholary O, Cornely OA, Anaissie E (2014) Improvement in the outcome of invasive fusariosis in the last decade. *Clin Microbiol Infect* 20:580–585

Nucci F, Nouer SA, Capone D, Anaissie E, Nucci M (2015) Fusariosis. *Semin Respir Crit Care Med* 36:706–714

Nucci M, Barreiros G, Akiti T, Anaissie E, Nouer SA (2021) Invasive fusariosis in patients with hematologic diseases. *J Fungi (Basel)* 7:815

Odio CD, Marciano BE, Galgiani JN, Holland SM (2017) Risk factors for disseminated coccidioidomycosis, United States. *Emerg Infect Dis* 23:308–311

O'donnell K, Rooney AP, Proctor RH, Brown DW, Mccormick SP, Ward TJ, Frandsen RJ, Lysoe E, Rehner SA, Aoki T, Robert VA, Crous PW, Groenewald JZ, Kang S, Geiser DM (2013)

Phylogenetic analyses of RPB1 and RPB2 support a middle Cretaceous origin for a clade comprising all agriculturally and medically important fusaria. *Fungal Genet Biol* 52:20–31
Omrani AS, Almaghrabi RS (2017) Complications of hematopoietic stem transplantation: fungal infections. *Hematol Oncol Stem Cell Ther* 10:239–244
Otu A, Kosmidis C, Mathioudakis AG, Ibe C, Denning DW (2023) The clinical spectrum of aspergillosis in chronic obstructive pulmonary disease. *Infection*:1–17
Panackal AA, Li H, Kontoyiannis DP, Mori M, Perego CA, Boeckh M, Marr KA (2010) Geoclimatic influences on invasive aspergillosis after hematopoietic stem cell transplantation. *Clin Infect Dis* 50:1588–1597
Pappas PG, Alexander BD, Andes DR, Hadley S, Kauffman CA, Freifeld A, Anaissie EJ, Brumble LM, Herwaldt L, Ito J, Kontoyiannis DP, Lyon GM, Marr KA, Morrison VA, Park BJ, Patterson TF, Perl TM, Oster RA, Schuster MG, Walker R, Walsh TJ, Wannemuehler KA, Chiller TM (2010) Invasive fungal infections among organ transplant recipients: results of the transplant-associated infection surveillance network (TRANSNET). *Clinical Infectious Diseases* 50:1101–1111
Pappas PG, Lionakis MS, Arendrup MC, Ostrosky-Zeichner L, Kullberg BJ (2018) Invasive candidiasis. *Nat Rev Dis Primers* 4:18026
Parslow BY, Thornton CR (2022) Continuing shifts in epidemiology and antifungal susceptibility highlight the need for improved disease management of invasive candidiasis. *Microorganisms* 10
Pasquier E, Kunda J, de Beaudrap P, Loyse A, Temfack E, Molloy SF, Harrison TS, Lortholary O (2018) Long-term mortality and disability in cryptococcal meningitis: a systematic literature review. *Clin Infect Dis* 66:1122–1132
Pecanha PM, Pecanha-Pietrobom PM, Grao-Velloso TR, Rosa Junior M, Falqueto A, Goncalves SS (2022) Paracoccidioidomycosis: what we know and what is new in epidemiology, diagnosis, and treatment. *J Fungi (Basel)* 8
Perfect JR (2015) Cryptococcosis (*Cryptococcus neoformans* and *Cryptococcus gattii*). In: Bennett JE, Dolin R, Blaser MJ (eds) Mandell, Douglas, and Bennett's principles and practice of infectious diseases, 8th edn. Philadelphia, Elsevier/Saunders
Perfect JR (2020) Cryptococcosis (*Cryptococcus neoformans* and *Cryptococcus gattii*). In: Bennett JE, Dolin R, Blaser MJ (eds) Mandell, Douglas, and Bennett's principles and practice of infectious diseases, 9th edn. Elsevier, Philadelphia
Perlin DS, Rautemaa-Richardson R, Alastruey-Izquierdo A (2017) The global problem of antifungal resistance: prevalence, mechanisms, and management. *Lancet Infect Dis* 17:e383–e392
Petrikkos G, Skiada A, Lortholary O, Roilides E, Walsh TJ, Kontoyiannis DP (2012) Epidemiology and clinical manifestations of mucormycosis. *Clin Infect Dis* 54(Suppl 1):S23–S34
Pfaller MA, Diekema DJ (2007) Epidemiology of invasive candidiasis: a persistent public health problem. *Clin Microbiol Rev* 20:133–163
Prakash H, Chakrabarti A (2021) Epidemiology of mucormycosis in India. *Microorganisms* 9:523
Prakash S, Kumar A (2023) Mucormycosis threats: a systemic review. *J Basic Microbiol* 63:119–127
Pruksaphon K, Nosanchuk JD, Ratanabanangkoon K, Youngchim S (2022) *Talaromyces marneffei* infection: virulence, intracellular lifestyle and host defense mechanisms. *J Fungi (Basel)* 8:200
Qin Y, Huang X, Chen H, Liu X, Li Y, Hou J, Li A, Yan X, Chen Y (2020) Burden of *Talaromyces marneffei* infection in people living with HIV/AIDS in Asia during ART era: a systematic review and meta-analysis. *BMC Infect Dis* 20:551
Quindos G, Marcos-Arias C, San-Millan R, Mateo E, Eraso E (2018) The continuous changes in the aetiology and epidemiology of invasive candidiasis: from familiar *Candida albicans* to multiresistant *Candida auris*. *Int Microbiol* 21:107–119
Rajasingham R, Smith RM, Park BJ, Jarvis JN, Govender NP, Chiller TM, Denning DW, Loyse A, Boulware DR (2017) Global burden of disease of HIV-associated cryptococcal meningitis: an updated analysis. *Lancet Infect Dis* 17:873–881
Ramirez-Garcia A, Pellon A, Rementeria A, Buldain I, Barreto-Bergter E, Rollin-Pinheiro R, de Meirelles JV, Xisto M, Ranque S, Havlicek V, Vandeputte P, Govic YL, Bouchara JP,

Giraud S, Chen S, Rainer J, Alastruey-Izquierdo A, Martin-Gomez MT, López-Soria LM, Peman J, Schwarz C, Bernhardt A, Tintelnot K, Capilla J, Martin-Vicente A, Cano-Lira J, Nagl M, Lackner M, Irinyi L, Meyer W, de Hoog S, Hernando FL (2018) *Scedosporium* and *Lomentospora*: an updated overview of underrated opportunists. *Med Mycol* 56:102–125

Rees JR, Pinner RW, Hajjeh RA, Brandt ME, Reingold AL (1998) The epidemiological features of invasive mycotic infections in the San Francisco Bay area, 1992–1993: results of population-based laboratory active surveillance. *Clin Infect Dis* 27:1138–1147

Reid G, Lynch JP 3rd, Fishbein MC, Clark NM (2020) Mucormycosis. *Semin Respir Crit Care Med* 41:99–114

Respaldiza N, Medrano FJ, Medrano AC, Varela JM, de LA Horra C, Montes-Cano M, Ferrer S, Wichmann I, Gargallo-Viola D, Calderon EJ (2004) High seroprevalence of Pneumocystis infection in Spanish children. *Clin Microbiol Infect* 10:1029–1031

Richardson MD (2005) Changing patterns and trends in systemic fungal infections. *J Antimicrob Chemother* 56(Suppl 1):i5–i11

Richardson M, Lass-Florl C (2008) Changing epidemiology of systemic fungal infections. *Clin Microbiol Infect* 14(Suppl 4):5–24

Rijnders BJA, Schauwvlieghe A, Wauters J (2020) Influenza-associated pulmonary aspergillosis: a local or global lethal combination? *Clin Infect Dis* 71:1764–1767

Roden MM, Zaoutis TE, Buchanan WL, Knudsen TA, Sarkisova TA, Schaufele RL, Sein M, Sein T, Chiou CC, Chu JH, Kontoyiannis DP, Walsh TJ (2005) Epidemiology and outcome of zygomycosis: a review of 929 reported cases. *Clin Infect Dis* 41:634–653

Rodrigues AM, Hagen F, Puccia R, Hahn RC, de Camargo ZP (2023) *Paracoccidioides* and paracoccidioidomycosis in the 21st century. *Mycopathologia* 188(1-2):129–133

Rodriguez-Tudela JL, Berenguer J, Guarro J, Kantarcioglu AS, Horre R, de Hoog GS, Cuenca-Estrella M (2009) Epidemiology and outcome of *Scedosporium prolificans* infection, a review of 162 cases. *Med Mycol* 47:359–370

Ruhnke M (2019) Clinical syndromes: *Candida* and Candidosis. In: Presterl E (ed) *Clinically Relevant Mycoses*, 1st edn. Springer, New York, NY

Sabino R, Goncalves P, Martins Melo A, Simoes D, Oliveira M, Francisco M, Viegas C, Carvalho D, Martins C, Ferreira T, Toscano C, Simoes H, Verissimo C (2021) Trends on *Aspergillus* epidemiology-perspectives from a National Reference Laboratory Surveillance Program. *J Fungi (Basel)* 7:28

Saegeman V, Maertens J, Meersseman W, Spriet I, Verbeken E, Lagrou K (2010) Increasing incidence of mucormycosis in University Hospital, Belgium. *Emerg Infect Dis* 16:1456–1458

Samson RA, Visagie CM, Houbraken J, Hong SB, Hubka V, Klaassen CH, Perrone G, Seifert KA, Susca A, Tanney JB, Varga J, Kocsube S, Szigeti G, Yaguchi T, Frisvad JC (2014) Phylogeny, identification and nomenclature of the genus *Aspergillus*. *Stud Mycol* 78:141–173

Scantlebury CE, Pinchbeck GL, Loughnane P, Aklilu N, Ashine T, Stringer AP, Gordon L, Marshall M, Christley RM, Mccarthy AJ (2016) Development and evaluation of a molecular diagnostic method for rapid detection of *Histoplasma capsulatum* var. *farciminosum*, the causative agent of epizootic lymphangitis, in equine clinical samples. *J Clin Microbiol* 54:2990–2999

Scully MC, Baddley JW (2018) Epidemiology of histoplasmosis. Curr Fungal Infect Rep 12:51–58

Seagle EE, Williams SL, Chiller TM (2021) Recent trends in the epidemiology of fungal infections. *Infect Dis Clin North Am* 35:237–260

Sharma M, Chakrabarti A (2023) Candidiasis and other emerging yeasts. *Curr Fungal Infect Rep*:1–10

Shpitzer T, Keller N, Wolf M, Goldschmied-Reouven A, Bahar G, Bahar I, Kronenberg J, Feinmesser R, Talmi YP (2005) Seasonal variations in rhino-cerebral *Mucor* infection. *Ann Otol Rhinol Laryngol* 114:695–698

Silva JT, Torre-Cisneros J, Aguado JM (2018) Invasive aspergillosis in solid organ transplantation. *Rev Iberoam Micol* 35:206–209

Singh AK, Singh R, Joshi SR, Misra A (2021) Mucormycosis in COVID-19: a systematic review of cases reported worldwide and in India. *Diabetes Metab Syndr* 15:102146

Sivagnanam S, Sengupta DJ, Hoogestraat D, Jain R, Stednick Z, Fredricks DN, Hendrie P, Whimbey E, Podczervinski ST, Krantz EM, Duchin JS, Pergam SA (2017) Seasonal clustering of sinopulmonary mucormycosis in patients with hematologic malignancies at a large comprehensive cancer center. *Antimicrob Resist Infect Control* 6:123

Skiada A, Pavleas I, Drogari-Apiranthitou M (2020) Epidemiology and diagnosis of mucormycosis: an update. *J Fungi (Basel)* 6:265

Skiada A, Drogari-Apiranthitou M, Pavleas I, Daikou E, Petrikkos G (2022) Global cutaneous mucormycosis: a systematic review. *J Fungi (Basel)* 8:194

Sow D, Ndiaye M, Sarr L, Kante MD, Ly F, Diousse P, Faye BT, Gaye AM, Sokhna C, Ranque S, Faye B (2020) Mycetoma epidemiology, diagnosis management, and outcome in three hospital centres in Senegal from 2008 to 2018. *PLoS One* 15:e0231871

Specjalski K, Kita K, Kuziemski K, Tokarska B, Gorska L, Szade J, Sieminska A, Chelminska M, Jassem E (2019) Histoplasmosis in an elderly polish tourist – a case report. *BMC Pulm Med* 19:150

Spellberg B (2012) Gastrointestinal mucormycosis: an evolving disease. *Gastroenterol Hepatol (N Y)* 8:140–142

Steinbrink JM, Miceli MH (2021) Mucormycosis. *Infect Dis Clin North Am* 35:435–452

Strollo S, Lionakis MS, Adjemian J, Steiner CA, Prevots DR (2017) Epidemiology of hospitalizations associated with invasive candidiasis, United States, 2002–2012. *Emerg Infect Dis* 23:7–13

Takashima M, Sugita T (2022) Taxonomy of pathogenic yeasts *Candida*, *Cryptococcus*, *Malassezia*, and *Trichosporon*. *Med Mycol J* 63:119–132

Talamantes J, Behseta S, Zender CS (2007) Statistical modeling of valley fever data in Kern County, California. *Int J Biometeorol* 51:307–313

Taylor JW, Barker BM (2019) The endozoan, small-mammal reservoir hypothesis and the life cycle of *Coccidioides* species. *Med Mycol* 57:S16–S20

Taylor ML, Reyes-Montes MDR, Estrada-Barcenas DA, Zancope-Oliveira RM, Rodriguez-Arellanes G, Ramirez JA (2022) Considerations about the geographic distribution of *Histoplasma* species. *Appl Environ Microbiol* 88:e0201021

Teixeira Mde M, Patane JS, Taylor ML, Gomez BL, Theodoro RC, de Hoog S, Engelthaler DM, Zancope-Oliveira RM, Felipe MS, Barker BM (2016) Worldwide phylogenetic distributions and population dynamics of the genus Histoplasma. *PLoS Negl Trop Dis* 10:e0004732

Thompson GR 3rd, Young JH (2021) *Aspergillus* infections. *N Engl J Med* 385:1496–1509

Tortorano AM, Prigitano A, Esposto MC, Arsic Arsenijevic V, Kolarovic J, Ivanovic D, Paripovic L, Klingspor L, Nordøy I, Hamal P, Arikan Akdagli S, Ossi C, Grancini A, Cavanna C, Lo Cascio G, Scarparo C, Candoni A, Caira M, Drogari Apiranthitou M (2014) European Confederation of Medical Mycology (ECMM) epidemiological survey on invasive infections due to *Fusarium* species in Europe. *Eur J Clin Microbiol Infect Dis* 33:1623–1630

Trifilio SM, Bennett CL, Yarnold PR, Mckoy JM, Parada J, Mehta J, Chamilos G, Palella F, Kennedy L, Mullane K, Tallman MS, Evens A, Scheetz MH, Blum W, Kontoyiannis DP (2007) Breakthrough zygomycosis after voriconazole administration among patients with hematologic malignancies who receive hematopoietic stem-cell transplants or intensive chemotherapy. *Bone Marrow Transplant* 39:425–429

Tsang CC, Tang JYM, Ye H, Xing F, Lo SKF, Xiao C, Han L, Wu AKL, Ngan AHY, Law KC, To YN, Sze DTT, Chim OHY, Hui TWS, Leung CWH, Zhu T, Yao C, Tse BPH, Lau SKP, Woo PCY (2020) Rare/cryptic *Aspergillus* species infections and importance of antifungal susceptibility testing. *Mycoses* 63:1283–1298

Turissini DA, Gomez OM, Teixeira MM, Mcewen JG, Matute DR (2017) Species boundaries in the human pathogen *Paracoccidioides*. *Fungal Genet Biol* 106:9–25

Turner SA, Butler G (2014) The *Candida* pathogenic species complex. *Cold Spring Harb Perspect Med* 4:a019778

Valdez AF, Miranda DZ, Guimaraes AJ, Nimrichter L, Nosanchuk JD (2022) Pathogenicity & virulence of *Histoplasma capsulatum* – a multifaceted organism adapted to intracellular environments. *Virulence* 13:1900–1919

Vallabhaneni S, Mody RK, Walker T, Chiller T (2016) The global burden of fungal diseases. *Infect Dis Clin North Am* 30:1–11
van de Peppel RJ, Visser LG, Dekkers OM, de Boer MGJ (2018) The burden of invasive aspergillosis in patients with haematological malignancy: a meta-analysis and systematic review. *J Infect* 76:550–562
van de Sande WW (2013) Global burden of human mycetoma: a systematic review and meta-analysis. *PLoS Negl Trop Dis* 7:e2550
van Diepeningen AD, Al-Hatmi AMS, Brankovics B, de Hoog GS (2014) Taxonomy and clinical spectra of *Fusarium* species: where do we stand in 2014? *Curr Clin Microbiol Rep* 1:10–18
van Elden LJ, Walenkamp AM, Lipovsky MM, Reiss P, Meis JF, de Marie S, Dankert J, Hoepelman AI (2000) Declining number of patients with cryptococcosis in the Netherlands in the era of highly active antiretroviral therapy. *Aids* 14:2787–2788
Vargas SL, Hughes WT, Santolaya ME, Ulloa AV, Ponce CA, Cabrera CE, Cumsille F, Gigliotti F (2001) Search for primary infection by *Pneumocystis carinii* in a cohort of normal, healthy infants. *Clin Infect Dis* 32:855–861
Vera C, Rueda ZV (2021) Transmission and colonization of *Pneumocystis jirovecii*. *J Fungi (Basel)* 7:979
Verma P, Jha A (2019) Mycetoma: reviewing a neglected disease. *Clin Exp Dermatol* 44:123–129
Verweij PE, Rijnders BJA, Bruggemann RJM, Azoulay E, Bassetti M, Blot S, Calandra T, Clancy CJ, Cornely OA, Chiller T, Depuydt P, Giacobbe DR, Janssen NAF, Kullberg BJ, Lagrou K, Lass-Florl C, Lewis RE, Liu PW, Lortholary O, Maertens J, Martin-Loeches I, Nguyen MH, Patterson TF, Rogers TR, Schouten JA, Spriet I, Vanderbeke L, Wauters J, van de Veerdonk FL (2020) Review of influenza-associated pulmonary aspergillosis in ICU patients and proposal for a case definition: an expert opinion. *Intensive Care Med* 46:1524–1535
Vilela R, Huebner M, Vilela C, Vilela G, Pettersen B, Oliveira C, Mendoza L (2021) The taxonomy of two uncultivated fungal mammalian pathogens is revealed through phylogeny and population genetic analyses. *Sci Rep* 11:18119
Vincent JL, Rello J, Marshall J, Silva E, Anzueto A, Martin CD, Moreno R, Lipman J, Gomersall C, Sakr Y, Reinhart K, Investigators EIGO (2009) International study of the prevalence and outcomes of infection in intensive care units. *Jama* 302:2323–2329
Walzer PD (2013) The ecology of pneumocystis: perspectives, personal recollections, and future research opportunities. *J Eukaryot Microbiol* 60:634–645
Wang F, Han R, Chen S (2023) An overlooked and underrated endemic mycosis-talaromycosis and the pathogenic fungus *Talaromyces marneffei*. *Clin Microbiol Rev* e0005122
Webb BJ, Ferraro JP, Rea S, Kaufusi S, Goodman BE, Spalding J (2018) Epidemiology and clinical features of invasive fungal infection in a US Health Care Network. *Open Forum Infect Dis* 5:ofy187
Wei HY, Liang WJ, Li B, Wei LY, Jiang AQ, Chen WD, Guo PH, Xu J (2021) Clinical characteristics and risk factors of *Talaromyces marneffei* infection in human immunodeficiency virus-negative patients: a retrospective observational study. *World J Emerg Med* 12:281–286
Weissenbacher-Lang C, Blasi B, Bauer P, Binanti D, Bittermann K, Ergin L, Högler C, Högler T, Klier M, Matt J, Nedorost N, Silvestri S, Stixenberger D, Ma L, Cissé OH, Kovacs JA, Desvars-Larrive A, Posautz A, Weissenböck H (2023) Detection of *Pneumocystis* and morphological description of fungal distribution and severity of infection in thirty-six mammal species. *J Fungi* 9:220
Weyant RB, Kabbani D, Doucette K, Lau C, Cervera C (2021) *Pneumocystis jirovecii*: a review with a focus on prevention and treatment. *Expert Opin Pharmacother* 22:1579–1592
Wheat LJ, Azar MM, Bahr NC, Spec A, Relich RF, Hage C (2016) Histoplasmosis. *Infect Dis Clin North Am* 30:207–227
White PL, Price JS, Backx M (2018) Therapy and management of *Pneumocystis jirovecii* infection. *J Fungi (Basel)* 4:127
WHO (2022) *WHO fungal priority pathogens list to guide research, development and public health action*. World Health Organization, Geneva

Williams SL, Chiller T (2022) Update on the epidemiology, diagnosis, and treatment of coccidioidomycosis. *J Fungi (Basel)* 8:666

Williamson PR, Jarvis JN, Panackal AA, Fisher MC, Molloy SF, Loyse A, Harrison TS (2017) Cryptococcal meningitis: epidemiology, immunology, diagnosis and therapy. *Nat Rev Neurol* 13:13–24

Wisplinghoff H, Bischoff T, Tallent SM, Seifert H, Wenzel RP, Edmond MB (2004) Nosocomial bloodstream infections in US hospitals: analysis of 24,179 cases from a prospective nationwide surveillance study. *Clin Infect Dis* 39:309–317

Woods JP (2016) Revisiting old friends: developments in understanding *Histoplasma capsulatum* pathogenesis. *J Microbiol* 54:265–276

Xu L, Chen X, Yang X, Jiang H, Wang J, Chen S, Xu J (2023) Disseminated *Talaromyces marneffei* infection after renal transplantation: a case report and literature review. *Front Cell Infect Microbiol* 13:1115268

Yu X, Cai X, Xu X, Zhang L, Huang X, Wang L, Chen Y (2018) Fungemia caused by *Penicillium marneffei* in an immunocompetent patient with COPD: a unique case report. *Medicine (Baltimore)* 97:e9658

Zavala S, Baddley JW (2020) Cryptococcosis. *Semin Respir Crit Care Med* 41:69–79

Zijlstra EE, van de Sande WWJ, Welsh O, Mahgoub ES, Goodfellow M, Fahal AH (2016) Mycetoma: a unique neglected tropical disease. *Lancet Infect Dis* 16:100–112

# The Role of the Immune System Against Neuromycotic Infections

**Vanielle A. do Nascimento Vicente, Ana Flávia Tostes, and Iane Carvalho Shieh**

## Introduction

Although they have completely different functions, the eyes, pregnant uterus, testicles, and brain have one characteristic in common: immunological privilege (Benhar et al. 2012; Forrester et al. 2018). This term, designated by Peter Medawar in the 1940s, served as a possible explanation for the absence of immunoreactivity observed in tissues transplanted in the brain and anterior chamber of the eye (Taylor and Kaplan 2010; Medawar 1948), suggesting that these structures are resistant to the action of foreign antigens because they lack a more reactive immune system. The experiments that demonstrated the almost absent immunological response in the nervous system involved injection of dye into the periphery, which in turn did not stain the brain and other studies in which the use of grafts in brain tissue did not cause intense immunological responses as was observed in other peripheral tissues. Furthermore, the presence of a specific cell type responsible for the immune response, microglia, corroborated this assumption (Solaro et al. 2022; Castellani et al. 1979).

This concept, which has become controversial, was used to explain the immunological adaptation that these organs, vital to the life of mammals, acquired to protect themselves from damage caused by inflammatory action in response to pathogens or foreign antigens in general. In the brain, in particular, this characteristic is very important given the little regenerative capacity that the cells in this tissue have, which could lead to compromised function (Hong and Van Kaer 1999; Galea et al.

V. A. do Nascimento Vicente (✉) · A. F. Tostes
University of São Paulo (USP), Institute of Biomedical Sciences,
Department of Physiology and Biophysics, São Paulo, SP, Brazil
e-mail: vaniellevicente@usp.br

I. C. Shieh
DASA, Technical Operations Center, Barueri, SP, Brazil

 
A. Gomes Rodrigues, A. Gupta (eds.), *Fungal Infections*, Fungal Biology,
https://doi.org/10.1007/978-3-032-06014-3_7

2007). Next, the physiological components of immune privilege at the cellular and molecular level will be described.

## *Physiological Components That Contributed to the Theory of Immune Privilege*

Some of the physiological factors studied to explain immune privilege mentioned here are related to two structures: the brain and the eyes, as the second is practically an extension of the first (London et al. 2012). We will start with MHC (Major Histocompatibility Complex) class I, which, although expressed in practically all nucleated cells in the body, is poorly expressed in neural and retinal cells (Niederkorn 2006; Kristensson 2011). MHC molecules have the function of associating with antigenic peptides that will be recognized by specific T lymphocytes, where MHC class I is presented to CD8+ T lymphocytes, cytotoxic cells that lead to apoptosis (Elmer and McAllister 2012; Wieczorek et al. 2017). The occurrence of cell death in the eye and brain would affect visual and neural activity, after all, these cell types have little or no regenerative capacity (Weishaupt and Zhang 2016). Therefore, the low expression of MHC class I prevents the action of cytotoxic T lymphocytes on neural cells, preventing cell death that could be mediated by lymphoid cells.

Little is known about the presence of anti-inflammatory and immunosuppressive elements solubilized in the cerebral interstitium; however, there is the possibility that gangliosides may inhibit the expression of MHC classes I and II, preventing immunoreactivity (Niederkorn 2006). Gangliosides are a type of glycosphingolipid that act as synaptic support and have been implicated in the plasticity of the nervous system (Palmano et al. 2015). Furthermore, they play an important role as mediators of the immune response and intestinal inflammation (Rueda 2013). This is important given that there are currently countless studies on the interaction between brain and intestine, known as the brain-gut axis, which culminates in the "dialogue" on both sides between these two organs of the human body (Carabotti et al. 2015), which has been linked to the development of neurological disorders such as Parkinson's and Alzheimer's (Barbosa and Barbosa 2020).

## *Molecules in the Cell Membrane That Affect Immune Privilege*

The Fas ligand (FasL), which can also be called CD59 ligand (CD59L), is widely expressed in the central nervous system, both in neural cells and in the vascular endothelium. This cytokine is also a constituent of immune privilege in the brain (Niederkorn 2006). But after all, how does this protein protect the brain and eyes from exacerbated immunological and inflammatory reactions?

Fas/CD59, a member of the tumor necrosis factor (TNF) family, is a protein that regulates the complement system, a set of plasma-soluble proteins that initiate a cascade of events that culminate in the maintenance of homeostasis through the elimination of infectious agents (Merle et al. 2015). The Fas protein inhibits the membrane attack complex (MAC), which is formed by complement proteins whose purpose is to form pores in the cell membrane, leading to lysis and, consequently, its death. Therefore, Fas/CD59 limits cell lysis caused by MAC, protecting cells from injury by complement cells. In fact, the Fas/CD59 protein is critical for this process as it is the only one with MAC inhibitory function in human cells (Shaz 2009; Couves et al. 2023; Alegretti et al. 2009).

In this way, central nervous systems (CNS) cells and vascular endothelium that express Fas/CD59 on the membrane are protected from inflammation mediated by innate or adaptive immunity factors, in particular, complement system proteins. Furthermore, studies show that deficits in FasL proteins and another protein, CD55, are associated with worsening conditions in patients with autoimmune diseases (Alegretti et al. 2009).

For a long time, the concept of immune privilege predominated in the field of neuroimmunology as an unquestionable truth, even if other works suggested otherwise. However, frequent technological and scientific advances have allowed greater discoveries in this field and the concept of a brain fully privileged with immunity has been challenged, suggesting that the nervous system has a guardian immune system that resides in structures that surround it (Castellani et al. 1979; Hubbard and Binder 2016; Kwon 2022). Next, recent discoveries on how the nervous system interacts with the immune system, which has a highly specialized immunological complexity, will be described.

## Physical and Immunological Barriers in the CNS: Recent Updates on Neuroimmunology

The brain is protected, from an anatomical point of view, by the skull, meninges, blood-cerebrospinal fluid barrier (BCSFB) and blood-brain barrier (BBB). Recent studies show that some of these structures serve not only as a physical barrier but also as an immunological protection, allowing immune system cells to migrate from adjacent portions of the central nervous system directly to the brain parenchyma during infections (Solaro et al. 2022; Castellani et al. 1979). We will describe new discoveries about the protection of the nervous system below.

## *Protection Mediated by the Meninges and Bone Marrow of the Skull*

The structures of the nervous system are surrounded by membranes called meninges, which are formed by three named layers, from the outermost to the innermost portion, such as the dura mater, arachnoid mater, and pia mater. The most superficial of them, the dura mater, is formed by dense connective tissue and is attached to the inner surface of the skull (Ghannam and Al Kharazi 2023). This anatomical arrangement does not occur by chance, as recent work suggests that there are channels that connect the bone marrow of the skull with the dura mater that allow the migration of myeloid cells to the CNS, promoting mechanisms of development of immature cells similar to those observed in the bone marrow. These cells, when they do not penetrate the brain tissue, act as guards in strategic locations such as the transverse and sagittal sinus (Solaro et al. 2022; Castellani et al. 1979), channels that drain blood and CSF to the internal jugular veins and from there to the vascular system (Castellani et al. 1979; Massrey et al. 2018).

Herisson and colleagues, using a technique that allowed marking the plasma membrane of bone marrow cells from the skull and tibia (rodents), determining the origin of the white cells found in tissues that had previously been subjected to inflammation, observed that the cells of the myeloid lineage, especially neutrophils, moved from the skull bone marrow towards the dura mater, through microchannels that connect the two structures (Herisson et al. 2018). This was one of the pioneering works on the discovery of these microchannels that connect the skull bones with the dura mater.

The work of Herisson aimed to study this migration in pathological situations, while the group of Cugurra demonstrated that this movement of skull bone marrow cells not only occurs in disease situations but also suggested that they are important for maintaining the homeostasis. Furthermore, it has been demonstrated that this cell type presents a distinct immunological phenotype when compared to those originating from the blood, as they are less harmful to nervous tissue. Thus, immune cells that reside in the dura mater have a more regulatory role that aims to mitigate the inflammatory response in neural structures (Cugurra et al. 2021).

It is currently known that cells of both innate and adaptive immunity reside in the dura mater (Su et al. 2023). Brioschi et al. studied the mechanisms of B cell migration in normal situations and during aging. B cells originating from the calvaria, that is, skull bones, travel through the same channels as innate immunity cells, and undergo the maturation process in the meningeal space, in the dura mater. It is believed that these cells have the particularity of preventing the production of immunoglobulins that have affinity to specific CNS epitopes, preventing exacerbated immunological reactions that could cause damage to this tissue. The same work observed that during aging, reactive B cells from the circulation migrate to the CNS, which may signal a threat to its structure and function (Brioschi et al. 2021).

Still in the dura mater, mural and endothelial cells, which provide support and nutrition for the nervous system, help direct antigens to T lymphocytes, through the action of chemokines. Furthermore, these cells allow the recruitment and survival of leukocytes through the release of some factors that allow this, such as type I insulin-like growth factor (IGF-I). All these data corroborate the hypothesis that the meninges, in addition to protecting against physical damage, can provide immunological surveillance provided by the cells residing there (Castellani et al. 1979).

## *Protection Mediated by the Blood-CSF Barrier*

The barrier between blood and cerebrospinal fluid is provided by cells of the choroid plexus, a type of epithelium that is responsible for the production of cerebrospinal fluid that runs throughout the ventricular system, being present in each ventricle of the brain (Castellani et al. 1979; Liddelow 2015). Many cells of the immune system are present in this environment, especially macrophages, which, due to their anatomical location, are in contact with the cerebrospinal fluid and act as immunological guards. In addition to these factors, the choroid plexus presents antigen-presenting cells (APCs) that express MHC class II, serving as a mechanism for attracting CD4+ T lymphocytes, which recognize antigens associated with this type of MHC (Castellani et al. 1979; Holling et al. 2004). In this way, the choroid plexus acts as a physical barrier and also provides protection for the nervous system, a pathway dependent on latent immunological stimuli (Castellani et al. 1979).

Allied to the choroid plexus and the meninges is also the bone marrow of the skull. As already mentioned, there are channels that allow the interconnection between the cells produced in the skull bone marrow with the dura mater, there is also the importance of the fact that it acts as a place that hosts innate immunity cells, suggesting that the skull bone marrow has a potential deposit of immune cells that can migrate to the CNS (Castellani et al. 1979). Together, the meninges, choroid plexus and skull bone marrow ensure that the entry of immune system cells into the brain can bypass the blood-brain barrier (BBB). These findings challenge the notion of immune privilege widely adopted for years regarding the mechanisms that aim to protect the nervous system against infectious agents and correlate harmful immune responses with the damage observed in aging and autoimmune diseases (Solaro et al. 2022; Castellani et al. 1979). It is important to highlight that from a clinical point of view, these discoveries are interesting as they may provide new tools for the treatment of diseases that can affect the nervous system, including infections (Cugurra et al. 2021), such as those caused by fungi.

## The Blood-Brain Barrier

The brain, like other organs in the body, needs elements to maintain homeostasis. However, it is provided with barriers that limit the entry and exit of these factors (Upadhyay 2014). Through the blood, which travels throughout the body through blood vessels, the tissues receive nutrients and oxygen, and it is also through it that metabolic products and other waste, such as carbon dioxide, are eliminated. Blood vessels differ structurally, depending on the organ they supply (Daneman and Prat 2015a). That said, the vessels that irrigate the CNS are of the non-fenestrated type as the pores are absent, in addition, they have a series of particularities that enable the proper function of this system (Zlokovic 2008; Daneman 2012). The name given to the unique vascular structure of the CNS is the blood-brain barrier (BBB).

The nervous system is protected by the BBB, which is made up of endothelial cells, pericytes and astrocytes (Daneman and Prat 2015a; Abbott et al. 2010). This structure was discovered in the nineteenth century, through the observation that the injection of a dye into the bloodstream colored several organs, with the exception of the brain, which was only colored when the same dye was injected directly into the cerebrospinal fluid (Kadry et al. 2020; Young et al. 2019). Figure 1 summarizes the structures that form the blood-brain barrier.

They are part of the BBB, considered the only true physical barrier in the brain today (Castellani et al. 1979), endothelial cells, pericytes and astrocytes. Endothelial cells form the walls of blood vessels, and differ greatly from the same cell type found in other tissues due to their flat shape, the absence of fenestrations and the presence of gap junctions between them, which hinders the paracellular diffusion of solutes. Transcytosis, a type of macromolecule transport across the BBB, is reduced in these cells. The passage of nutrients through the BBB takes place through transporters or receptors, which move from the neural microenvironment to the blood circulation and vice versa, this explains the high level of mitochondria found in this cell. They share the basement membrane with another type of cell that forms the BBB, pericytes (Kadry et al. 2020; Young et al. 2019; Daneman and Prat 2015b; Abbott et al. 2010).

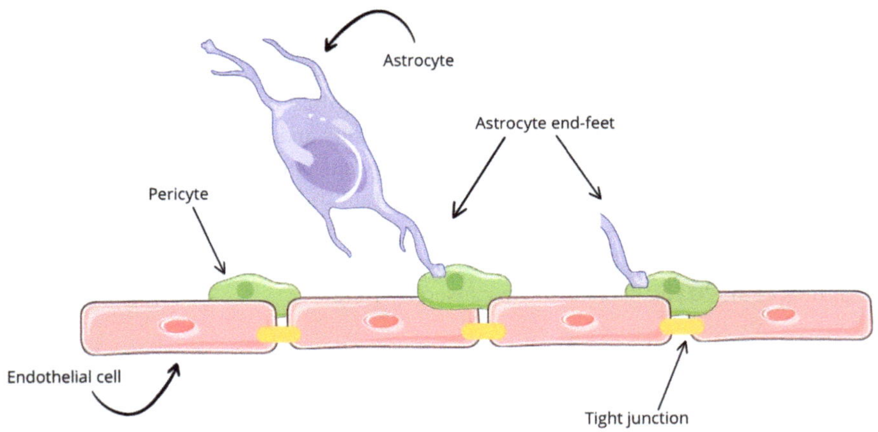


**Fig. 1** Schematic representation of the blood-brain barrier (BBB) and its components, astrocytes, endothelial cells, pericytes, and gap junctions. Some parts of the figure were created using images from Servier Medical Art, licensed under Creative Commons Attribution 4.0 (https://creativecommons.org/licenses/by/4.0/)

Pericytes, also known as mural cells, cover the wall of endothelial cells and provide structural support. They present contractility, similar to that observed in smooth muscles, therefore, controlling the blood flow that will regulate the synaptic activity of neurons. They are also considered relevant for angiogenesis and control of immune system cell infiltration in the CNS (Kadry et al. 2020; Young et al. 2019; Daneman and Prat 2015b).

Finally, the last element, the neurovascular unit, which comprises the previously mentioned cells together with neurons and microglia, are astrocytes. This type of glial cell is extremely important because it is the element that connects neural activity with vascular activity, given by the processes (feet) of the astrocytes that cover the endothelial cells and pericytes (Kadry et al. 2020; Young et al. 2019).

Under normal conditions, the BBB is capable of regulating the homeostasis of the neural microenvironment, and BBB dysfunction is related to some pathologies, such as Alzheimer's, Parkinson's, multiple sclerosis, diabetes, and stroke (Young et al. 2019; Daneman and Prat 2015b). Furthermore, infections, such as HIV are also related to damage to the BBB (Daneman 2012), especially due to the disruption of gap junctions present between endothelial cells, which has been observed in cases of HIV-associated encephalopathy (Dallasta et al. 1999). An important fact, since fungi are opportunistic organisms, that is, they tend to infect the nervous system mainly in immunocompromised patients (Gavito-Higuera et al. 2016a).

## Actions of the Immune System Against Fungal Infections

Mammals in general, unlike other living beings, are protected from fungal infections. One of the main agents for this protection is endothermy, the possibility of maintaining a constant and high temperature due to metabolic activity. Another characteristic that provides greater protection is the presence of a flexible immune system, especially in relation to adaptive immunity cells (Casadevall 2018; Woodring et al. 2023a). Other vertebrates, such as amphibians (frogs, tree frogs, toads), have not had the same luck, as they have been decimated by a fungus from the chytridiomycota phylum, the Bd fungus (*Batrachochytrium dendrobatidis*). This fungus leads to thickening of the skin of these animals, which affects their gas exchange, as these animals breathe cutaneously (O'Hanlon et al. 2018). In fact, fungi became a major medical problem after the advent of the HIV virus, reinforcing the statement that fungal infections particularly affect immunocompromised patients (Casadevall 2018), species of opportunistic pathogens such as *Aspergillus* sp., Pneumocystis and Cryptococcus being the most important in immunocompromised individuals (Woodring et al. 2023a).

The first innate barrier to be encountered by fungi is provided by the epithelial cells present in the mucous membranes, with the lung mucosa being one of the main entry routes for these microorganisms, as we frequently inhale spores present in the environment when we breathe. We can also add temperature as an innate barrier, which is associated with physical barriers, such as epithelial tissue, and chemical barriers (Woodring et al. 2023a). The next topics will include a description of the action of innate and adaptive immunity cells against fungal infections and the association of recent discoveries about neuroimmunology and its role in mycoses.

## *Receptors, Molecules, and Innate Immunity Cells*

Elements of innate immunity have the ability to recognize molecules that are part of the structure of certain microbes. These molecules are called pathogen-associated molecular patterns (PAMPs). In this way, each class of microorganism is capable of eliciting responses mediated through its PAMPs (Gow et al. 2017; Patin et al. 2019). Fungi have a cell wall formed especially by carbohydrates, such as chitin (polymer of N-acetyl-D-glucosamine), ß-glucans (glucose polymers connected by ß-glycosidic bonds) and mannan (mannose polymers), which surround the membrane plasma formed mainly by ergosterol, which, as previously mentioned, has a structure similar to cholesterol in animal cells (Stewart 2017a, b; Boynton and Ferneini 2016; Williams et al. 2016).

Pattern recognition receptors (PRRs) are present in the membrane of phagocytes and even epithelial cells, they are important and interact with PAMPs, leading to the activation of signaling cascades that culminate in the transcription and gene expression of several molecules, including chemokines and cytokines, which are important for recruiting more leukocytes to the site of infection and in inducing inflammatory activity (Muñoz-Wolf and Lavelle 2016).

The main PRRs of importance for mycotic infections are especially those that detect the presence of polysaccharides that form the fungal wall, such as the toll like receptors (TLRs), nucleotide oligomerization domains (NOD)-like receptors (NLRs), protease-activated receptors (PARs) and C-type lectin receptors (CLRs) (Williams et al. 2016; Bartemes and Kita 2018), which will be described in more detail below.

TLRs can be both cell surface (i.e., transmembrane) and intracellular receptors, of which 12 types are currently recognized. TLRs 1, 2, 4 and 6 are membrane receptors while TLRs 3 and 9 are present in the cytosol (i.e., intracellular). TLR1 recognizes ß-,glycans of fungal conidia and hyphae and components of Aspergillus fumigatus in humans. TLR2 is activated by ß-glycans from conidia and Coccidioides, by fungal hyphae, mannan from Candida albicans and molecules from Aspergillus fumigatus that are not yet known. TLR4 ligands can be mannans from Candida albicans and Cryptococcus neoformans and sugars from conidia. Whereas TLR6 recognizes Candida albicans. TLR3 and TLR9, which are intracellular, recognize double-stranded DNA from conidia and unmethylated CpG-rich DNA from fungi in general (Williams et al. 2016; Muñoz-Wolf and Lavelle 2016; Bartemes and Kita 2018).

In humans, 23 types of NLRs have been identified, of which the most studied is NLRP3. It can be activated by the hyphae of the fungus Aspergillus fumigatus, in addition, it activates a pathway that interacts with CLRs (Williams et al. 2016).

CLRs recognize fats, proteins and carbohydrates, the latter being the main components of the fungal wall. CLRs involved in the recognition of mycotic molecules are dectin-1, which is activated by ß-glucans from species of Candida, Aspergillus, Pneumocystis and Coccidioides, dectin-2, which interacts with fucose and mannose (*C. albicans* and Malassezia) as well as the receptor of mannose (MR), which is also activated by fucose. The last to be mentioned in this class of receptors is the dendritic cell-specific intercellular adhesion molecule-3-grabbing nointegrin (DC-SIGN), which is activated by mannan, α-mannosyl and α-fucosyl residues (Williams et al. 2016; Muñoz-Wolf and Lavelle 2016; Bartemes and Kita 2018).

Therefore, another group of receptors with an important role in the detection of fungal molecules are the PARs, which are part of the G protein-coupled receptor family, which generally lead to intracellular cascades involving the $Ca^{2+}$ ion, which result in the transcription of cytokines, chemokines and lipid mediators of inflammation, such as prostaglandins (Williams et al. 2016). A type of ascomycete, Alternaria, activates the PAR2 receptor and Aspergillus proteases suppress the chemokine CXCL10 through the same receptor (Williams et al. 2016; Bartemes and Kita 2018).

The interaction of PRRs with PAMPs activates cells, such as neutrophils, macrophages and dendritic cells (DCs), which are important for defense against mycoses (Woodring et al. 2023a; Jiang 2016). Phagocytic cell receptors bind to microbial components and lead to phagocytosis, and the internalized element within the neutrophil is called a phagosome. Inside the phagosome, neutrophils can release reactive oxygen species (ROS) and other elements that aim to completely destroy the pathogen (Naish et al. 2023). Another mechanism of pathogen destruction mediated by neutrophils is neutrophil extracellular traps (NETs), which aim to capture and destroy microorganisms, a mechanism mediated by the dectin-1 receptor. In fact, neutrophils represent the first line of defense against fungi, followed by macrophages (Woodring et al. 2023a; Jiang 2016).

Macrophages can enhance neutrophilic activity through the release of cytokines, chemokines (chemoattractant cytokines) to sites of *A. conidia* infection (Jiang 2016), for example, demonstrating the mutual activity that occurs between these two types of leukocytes. Furthermore, macrophages may have classic antimycotic activity (M1), or function with fungal reserves (M2), which is observed in cases of recurrent cryptococcosis (Woodring et al. 2023a).

It is not common for fungi to be intracellular pathogens, so the role of natural killer cells (NKs) is poorly described in fungal infections (Woodring et al. 2023a), although studies using cell cultures demonstrate the effectiveness of this cell type against some fungal agents (Jiang 2016). Finally, we will mention DCs, which are the bridge that connects the components of innate and adaptive immunity, as they are the main antigen-presenting cells for T lymphocytes (Woodring et al. 2023a; Jiang 2016; Wüthrich et al. 2012). Furthermore, DCs also have antifungal activity similar to other innate immunity cells, due to the presence of PRRs on their surface (Wüthrich et al. 2012; Woodring et al. 2023b).

## *The Role of Adaptive Immunity in Combating Mycoses*

Adaptive immunity has two highly important characteristics: specificity and memory. The first quality is related to planning the "attack" on pathogens, while the second provides efficiency in combating a second infection. The effector cells of adaptive immunity are B and T lymphocytes (Sun et al. 2023; Chi et al. 2024). However, the type of lymphocyte that plays the most significant role in combating fungal infections is the T lymphocyte, specifically the T helper (Th) or CD4+ T lymphocyte (Woodring et al. 2023a; Jiang 2016; Talbot et al. 2016). Th cells exhibit a certain complexity as they have several subtypes, and the types that are important in mycosis are named Th1, Th2, and Th17. They differ from each other due to the different proteins associated with their surface and the cytokines they secrete (Sun et al. 2023). Thus, each subtype of CD4+ T lymphocyte will elicit different pathways in combating fungal infections.

Th1 cells are activated by antigen-presenting cells and macrophages through the secretion of IL-12, and the response they trigger involves the secretion of cytokines IL-12, TNF-β, and IFNγ. IFNγ plays a fundamental role in combating intracellular pathogens, including fungi, by enhancing the activity of phagocytes and the opsonization promoted by B lymphocytes, ultimately leading to the death of infected cells. Furthermore, deficits in Th1 activity are associated with impaired responses in patients suffering from fungal infections (Woodring et al. 2023a; Bartemes and Kita 2018; Jiang 2016; Sun et al. 2023).

Th2 cells are activated by IL-4 and IL-13 and secrete IL-5. They are involved in humoral responses at the systemic level. IL-5 tends to limit the activity of Th1 and Th17, which is why Th2 activity is often associated with an impaired immune response to fungal infections, including intracellular fungi. Additionally, Th2 cells are linked to exacerbated allergic responses elicited by fungal antigens (Woodring et al. 2023a; Bartemes and Kita 2018; Jiang 2016). Although the effects of Th2 are not solely detrimental when related to mycoses, in infections by *Pneumocystis*, Th2 cells can promote the recruitment of eosinophils, which facilitate the pathogen's death (Woodring et al. 2023a).

Finally, Th17 cells are the primary cell type in adaptive immunity important for combating fungal infections, particularly in acute infections. They are stimulated by IL-1, IL-6, and TGFβ, and they secrete IL-17 and IL-22. IL-17 facilitates the action of phagocytes and inhibits the growth of intracellular pathogens, in addition to promoting the production of peptides with fungicidal activity (Woodring et al. 2023a; Bartemes and Kita 2018). Fungi such as *Pneumocystis carinii* and *Candida albicans* lead to a significant increase in Th17 cell activity due to the elevated levels of IL-23, a cytokine important for the differentiation of this cell type (Sun et al. 2023). Th17 cells primarily reside in mucosal tissues, and damage that affects their action can facilitate fungal infections in mucosal tissues, such as candidiasis (Jiang 2016).

There's indeed a delicate balance between Th1 and Th17 cells concerning fungal infections. In some cases, the same fungal species can trigger different types of helper T cells. For instance, C. albicans yeast can activate Th17, while pseudohyphae of the same fungus can activate Th1. Similarly, Aspergillus conidia may activate Th1 cells, while Aspergillus hyphae can activate Th17 cells. This variation depends on how these cells are activated by components of innate immunity and the specific pro-inflammatory cytokines that trigger different PRRs in innate immunity, leading to the activation of distinct cell types in acquired immunity (Woodring et al. 2023a; Bartemes and Kita 2018). Th17 cells indeed play a crucial role in the initial response to fungal infections, but both Th1 and Th17 cells are equally important in this function. Conversely, regulatory T cells (Treg) generally have the opposite effect and can dampen the immune response against fungal infections (Speakman et al. 2020).

## *The "Brain Borders Immunity" and Fungal Infections: New Insights*

The role of immune system cells that reside in the meninges, skull bone marrow, and structures of the perivascular space are the focus of recent studies in neuroimmunology, therefore, there is a lack of data relating these cells to the fight against fungal infections. Below, mere speculations that arise from these findings will be described.

As already mentioned, the meninges serve as a route for cells of the myeloid lineage produced by the bone marrow of the skull to travel to the nervous tissue (Drummond 2023), this may explain the neutrophilia observed in infections caused by *C. albicans* in the CNS, as until then the site of generation of these cells was not known.

Another interesting finding indicating the important role of cells residing at the edges of the CNS derives from postmortem studies in which strains of *Cryptococcus neoformans* were found in the intracellular environment of cells in the perivascular space, which from an anatomical point of view makes a lot of sense (Mohamed et al. 2022), which may be indicative of the immune role of these cells in antifungal activity. In fact, cryptococcal meningitis is the most common fungal infection in the CNS, especially in people whose CD4+ T cell activity is compromised in some way (Drummond 2023; Mohamed et al. 2022), which makes this species of fungal pathogen one of the most studied.

Therefore, a study demonstrated that microglia, in addition to serving as an intracellular reservoir of *C. neoformans*, did not demonstrate effective actions in combating this infection (Mohamed et al. 2023). This may indicate that other cell types, especially cells from the perivascular space, within the meninges and originating in the skull bone marrow, have a more effective role in combating this infection. In short, more studies are needed to unravel the role of guardian cells that live around the CNS in combating infections in general, especially fungal ones. Figure 2 shows an overview of the role of immune system against mycotic infections.

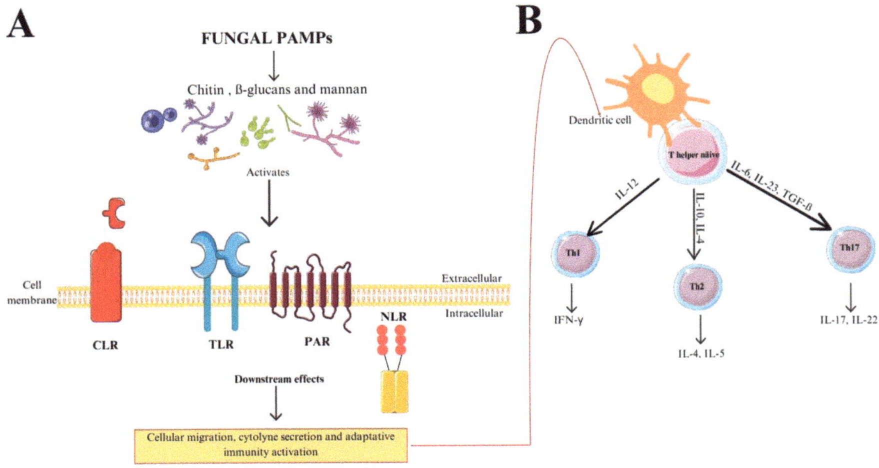


**Fig. 2** Overviewing the role of immune system in fungal infections. (**a**) The activation of pattern recognition receptors (PRRs) by fungal pathogen-associated molecular patterns (PAMPs) leads to a cascade of immune responses, including cellular migration, cytokine secretion. (**b**) Activation of adaptive immunity, particularly through the stimulation of antigen-presenting cells (APCs) that promote the differentiation of T helper (Th) cells. Adapted from Woodring et al. (2023a). Some parts of the figure were created using images from Servier Medical Art, licensed under Creative Commons Attribution 4.0 (https://creativecommons.org/licenses/by/4.0/)

## Fungal Infections and CNS: What Do We Know So Far?

Fungal infections in the CNS have become more common over the last two decades (Góralska et al. 2018a). Although it is estimated that there are 1.5 million fungal species, only about 700,000 have been formally described. Of the species described, 300 can be virulent for humans, and only 10–15% of these can influence the CNS (Sharma et al. 2012; Köhler et al. 2015). Clinically relevant fungi for fungal infections are yeasts from cosmopolitan species of Candida and Cryptococcus genera, filamentous fungi, characterized with branching hyphae, including light-colored moniliac fungi with septic hyphae, such as Aspergillus spp. and Fusarium spp., and non-septic fungi, such as Rhizopus, Rhizomucor, and Mucor, lastly, less common fungi like *Trichosporon* spp. These fungal species have a worldwide distribution and are common causes of fungal infections of the CNS (Murthy and Sundaram 2014a; Bongomin et al. 2017; McCarthy et al. 2014). These infections typically occur in immunocompromised patients and are often lethal, with challenging diagnosis and treatment (McCarthy et al. 2014; Liu et al. 2017; Stockamp and Thompson 2016) and penetration of the pathogen across the blood-brain barrier (BBB) is an essential step for CNS invasion (Góralska et al. 2018a).

Around 1.5 billion years ago, the common ancestor between fungi and animals separated, and molecular studies propose that fungi have greater phylogenetic proximity to animals than to plants (Woodring et al. 2023b; Asiegbu and Kovalchuk

2021). Normally, fungi have a cell membrane surrounded by a cell wall, in which the wall is formed by structures similar to the cholesterol found in the plasma membrane of animals, ergosterol and zymosterol (Carmona and Limper 2017). All of these together exemplify one of the difficulties of treating fungal infections, due to the similarities with mammalian cells (Boynton and Ferneini 2016), especially in immunosuppressed individuals, in individuals living with HIV, the virus affects CD4+ T cells, which are crucial for antifungal activity. This underscores the vulnerability of the immune system in people affected by HIV to fungal infections (Woodring et al. 2023a).

## *Fungal Infections in the CNS and the Interaction with the BBB*

For pathogens to cross the BBB, there are three mechanisms: transcellular migration, paracellular migration and the trojan horse mechanism (Koutsouras et al. 2017; Raman 2010). For the mentioned mechanisms, Candida and *Cryptococcus neoformans* are better understood. *Cryptococcus neoformans* is mechanically trapped in the cerebral vasculature and can cross the BBB by both direct and indirect mechanisms (Colombo and Rodrigues 2015). The direct pathway includes the passage of the BBB through endothelial cell transcytosis (Colombo and Rodrigues 2015; Casadevall 2010) while the indirect pathway includes transport within phagocytes like Trojan horse mechanisms (Colombo and Rodrigues 2015). For the processes of translocation of pathogenic proteins to occur through the BBB, it is necessary for interactions to exist between molecules of these proteins and cellular and transcellular mechanisms (Jong et al. 2008). The direct pathway includes the passage of the BBB through endothelial cell transcytosis (Colombo and Rodrigues 2015; Casadevall 2010) while the indirect pathway includes transport within phagocytes like Trojan horse mechanisms (Colombo and Rodrigues 2015). For the processes of translocation of pathogenic proteins to occur through the BBB, it is necessary for interactions to exist between molecules of these proteins and cellular and transcellular mechanisms (Jong et al. 2008), such as C. neoformans. Jong et al. showed that C-alpha protein kinase activation is required for the use of the transcellular mechanism by *C. neoformans* on cerebral microvascular endothelial cells (BMECs). In addition, the CPS1 gene is required for its adhesion to the CD44 surface protein of human BMECs, and the spread of the pathogen in the brain is controlled by the Iscs1 gene, which encodes an enzyme that hydrolyzes inositol. Once the pathogen is internalized by the phagocyte, it can actively manipulate the phagocyte to promote migration to the brain. The infected phagocyte reaches the brain and adheres to the luminal side of the cerebral capillaries and crosses the BBB, either paracellularly or transcellularly (Santiago-Tirado and Doering 2017).

In states of reduced immunity, the BBB becomes more permeable, allowing fungi to penetrate the brain more easily. The pathogens reach the brain parenchyma and proliferate, causing inflammation (Sharma et al. 2012). Thus, the invasion of pathogens in the brain is mainly associated with immunocompromised states, as the pathogenic factors need to overcome the effective barriers that surround the brain. The activation of nerve cells by fungal cells and the expression of immunodeficient and immunosuppressant cytokines and chemokines play a decisive role in the pathogenesis of systemic infections of the CNS (Sharma et al. 2012). Thus, CNS involvement occurs when invading fungi cross the BBB, brain, and subarachnoid spaces to reach the CNS (Koutsouras et al. 2017). Various disturbances of the BBB, such as trauma, surgery, or activation of microglia and cytokines, favor this process. The rate and extent of infection are influenced by the virulence of the fungus and the immune activity of the host (Koutsouras et al. 2017).

## *Fungal CNS Infections Diagnosis*

The diagnosis of invasive fungal infections is based on the combined interpretation of risk factors, clinical symptoms and imaging results according to EORTC/MSG criteria. In most cases, symptoms of CNS invasion are nonspecific and include headache, fever, seizures, muscle weakness, and progressive confusion. In addition, changes in mental status or focal neurological deficits may occur (Brumble et al. 2017; Gavito-Higuera et al. 2016b). For cryptococcosis, computed tomography (CT) imaging is usually negative, and focal lesions are seen in the course of filamentous fungal infections. On magnetic resonance imaging (MRI), nonspecific focal lesions, edema, or hemorrhagic lesions can be found (Massrey et al. 2018). However, CT and MRI techniques can only serve as additional aids in the diagnosis of fungal infections of the CNS (Gavito-Higuera et al. 2016b). CNS biopsies are considered very risky in critically ill patients, especially in a population of hematologic patients with low platelet counts or neutropenia (Arvanitis et al. 2014) However, biopsy allows obtaining samples from the CNS including those from the brain, meninges, and cerebrospinal fluid (CSF) or ventricular fluid (Arvanitis et al. 2014).

The direct pathway includes the passage of the BBB through endothelial cell transcytosis (Colombo and Rodrigues 2015; Casadevall 2010), while the indirect pathway includes transport within phagocytes like Trojan horse mechanisms (Colombo and Rodrigues 2015). For the processes of translocation of pathogenic proteins to occur through the BBC, it is necessary for there to be interactions between molecules of these proteins and cellular and transcellular mechanisms (Jong et al. 2008) such as *C. neoformans*. Jong et al. showed that C-alpha protein kinase activation is required for the use of the transcellular mechanism by *C. neoformans* on cerebral microvascular endothelial cells (BMECs). In addition, the CPS1 gene is required for its adhesion to the CD44 surface protein of human BMECs, and the spread of the pathogen in the brain is controlled by the Iscs1 gene, which encodes an enzyme that hydrolyzes inositol. Once the pathogen is internalized by the phagocyte, it can actively manipulate the phagocyte to promote migration to the brain. The infected phagocyte reaches the brain and adheres to the luminal side of the cerebral capillaries and crosses the BBB, either paracellularly or transcellularly (Santiago-Tirado and Doering 2017).

In states of reduced immunity, the BBB becomes more permeable, allowing fungi to penetrate the brain more easily. The pathogens reach the brain parenchyma and proliferate, causing inflammation (Sharma et al. 2012). Thus, the invasion of pathogens in the brain is mainly associated with immunocompromised states, as the pathogenic factors need to overcome the effective barriers that surround the brain. The activation of nerve cells by fungal cells and the expression of immunodeficient and immunosuppressant cytokines and chemokines play a decisive role in the pathogenesis of systemic infections of the CNS (Sharma et al. 2012). Thus, CNS involvement occurs when invading fungi cross the BBB and brain and subarachnoid spaces to reach the CNS (Koutsouras et al. 2017). Various disturbances of the BBB, such as trauma, surgery, or activation of microglia and cytokines, favor this process. The rate and extent of infection are influenced by the virulence of the fungus and the immune activity of the host (Koutsouras et al. 2017).

Methods based on optical brighteners, such as Calcofluor and or Blankophor allow direct examination and have high sensitivity and specificity for detecting fungal elements. Biopsy material stained with hematoxylin and eosin (HE), particularly with Gomori methenamine silver (GMS) or periodic acid-Schiff (PAS) is of great importance in the diagnosis of neuroinfections (Guarner and Brandt 2011; Patterson et al. 2016; Cornely et al. 2014).

## *CNS Fungal Infections Treatment*

Fungal neuroinfections are characterized by higher mortality rates and worse prognosis compared to viral, bacterial and parasitic infections (Raman 2010). Management of patients with fungal infections requires a multidisciplinary approach, early assessment, rapid diagnosis, and the use of appropriate and early therapy are crucial to help prevent an often fatal outcome (Su et al. 2023). For this, the choice of antifungal therapy depends on the fungistatic and fungicidal action of the therapeutic agent (Murthy and Sundaram 2014a; Raman 2010). Antifungal compounds vary in their physicochemical properties, which determine the extent of CNS penetration, spectrum of activity, and mode of action. Voriconazole, fluconazole, and flucytosine are antifungal agents that can more easily cross the BBB and reach the CNS, but fluconazole and flucytosine have a narrow spectrum of activity (Galgiani et al. 2016). In addition to pharmacological treatments, it is possible to remove the lesions surgically. Surgeries can be stereotaxic procedures, craniotomy, mycoplasty aneurysm surgery, and bypass surgery (Rajshekhar 2007). In most cases, the combination of surgical intervention and antifungal therapy increases the survival rate of patients (Galgiani et al. 2016).

Another commonly used treatment, especially in immunocompromised individuals, is the association of IFN$\gamma$ associated with antimycotics, which appears to improve the improvement in fungal infections due to the deficient role of CD4+ T lymphocytes (Drummond 2023).

## *The Principal Antifungal Agents Used to Treat Fungal Infections in the CNS*

The principal medicines used to treat fungal infections affecting the CNS are fluconazole, voriconazole, isavuconazole, itraconazole, posaconazole, amphotericin B, and flucytosine. With the exception of flucytosine, the mechanism of action of these medicines is to alter the structure of the cell membrane or interact with certain components of the fungal wall such as ergosterol, chitin, and β-glucans. Fluconazole, voriconazole, and isavuconazole promote inhibition of the 14 α-lanosterol demethylase of cytochrome P-450, leading to an accumulation of lanosterol, which, in turn, causes damage to the fungal cell membrane. Itraconazole and posaconazole act by inhibiting ergosterol synthesis in the fungal cell membrane. Amphotericin B acts by binding to ergosterol, which destabilizes the fungal cell membrane. Finally, flucytosine, which has a different mechanism of action to the other antifungal agents mentioned, acts by weakening the synthesis of fungal DNA and RNA (Góralska et al. 2018b; Bellmann and Smuszkiewicz 2017; Rodrigues 2020).

As for the spectrum of action of these medicines in the CNS, fluconazole has a spectrum for the treatment of Candida (except *Candida glabrata* and *Candida krusei*), Cryptococcus, Histoplasma, Blastomyces and Coccidioides infecions. Voriconazole's spectrum includes the treatment of Candida, Aspergillus, Fusarium and Scedosporium infections. Isavuconazole is used in the treatment of *Candida Aspergillus,* Mucor, Rhizopus, Rhizomucor, Fusarium, Sporothrix. Itraconazole is used as therapy against Aspergillus, Cryptococcus, Candida, Histoplasma, Paracoccidioides, Blastomyces, Sporothrix. The posaconazole has a spectrum of action against Aspergillus, Candida, Coccidioides, Fusarium, Rhizomucor, Mucor, Rhizopus. Amphotericin B is used as a therapeutic strategy against Mucor, Absidia, Aspergillus, Cryptococcus, Candida, Histoplasma, Blastomyces, Coccidioides, Paracoccidioides, Sporothrix and Flucytosine has a spectrum of action against Cryptococcus, Candida, *Cladophialophora Fonsecaea* and Phialophora (Góralska et al. 2018b; Bellmann and Smuszkiewicz 2017; Murthy and Sundaram 2014b; Boynton and Ferneini 2016).

## *The Biggest Challenges in the Treatment of Fungal Infections in the CNS*

As mentioned above, there are several challenges to be considered in relation to the therapeutic strategies employed in the management of fungal infections affecting the CNS. Among the greatest difficulties, we can mention the low permeation of the drug in the CNS, the reduced number of molecular targets and fungal resistance.

For example, the permeation of medicines into the CNS is determined by the private quality of the BBB in relation to the entry of substances (Mulvihill et al. 2020) The passage of medicines through the BBB is regulated by transporters, since the endothelial cells covering the cerebral blood vessel are not fenestrated. Another property involved in the penetration of medicines into the CNS is inflammation. During the inflammatory process, the endothelial cell junctions in the BBB are damaged, which usually promotes increased drug diffusion into the CNS. Finally, there are several physicochemical properties that determine the ability of medicines to cross the BBB. Among the physicochemical properties, the molecular weight (MW), the lipophilicity of the molecule and the ability to bind to plasma proteins can be mentioned (Wirth and Ishida 2020). Only after the drug has entered the CSN it accumulates to reach the concentration considered therapeutic. In addition to permeation capacity, the drug's spectrum of activity must also be considered. For example, voriconazole, fluconazole and flucytosine can easily penetrate the CNS, while itraconazole and posaconazole have lower permeation. However, it is important to note that fluconazole and flucytosine have a low spectrum of activity. Amphotericin B, on the other hand, has a low penetration rate in the CNS, but is highly effective in the treatment of fungal meningitis. In view of the toxicological characteristics of amphotericin B, voriconazole is recommended as primary therapy for CNS aspergillosis, while liposomal amphotericin B (L-AmB) is reserved for intolerant or refractory patients (Góralska et al. 2018b; Felton et al. 2014; Schwartz et al. 2018).

In addition to the difficulty of crossing the BBB, antifungal therapy is also hampered by the small number of selective molecular targets, since fungi are eukaryotic beings like their human host and most antifungal agents target ergosterol. These limited specific targets can lead to cross-resistance to all medicines that have the same target in common (Robbins et al. 2016).

## Resistance and Tolerance to Antifungal Agents in the CNS

Historically, the therapeutic strategy employed in the use of antifungal agents has been based on the four main classes of medicines: polyenes, azoles, echinocandins, and the pyrimidine analogue 5-flucytosine. However, these medicines stimulate a rapid response from the fungi and loss of treatment efficacy is commonly encountered. Failure to respond to antifungal agents can be caused by defects in the host's immune system, the pharmacokinetic and pharmacodynamic properties of antifungal agents, as well as the intrinsic characteristics of each fungus, which together can lead to tolerance and/or fungal resistance. Antifungal resistance is an evolutionary response given by the direct or indirect interaction of the medicine with the pathogen and is defined as the ability of fungi to grow when subjected to treatment with antifungal drugs at concentrations that stop growth and/or kill most fungi isolated from that species. In contrast, antifungal tolerance is the ability of fungi to grow when subjected to treatment with antifungal agentsat concentrations above the minimum inhibitory concentration (MIC) (Robbins et al. 2017; Fisher et al. 2022). In addition, there are more subtle responses that can have clinical significance and confer resistance/tolerance to the use of antifungal agents, such as heteroresistance, biofilm formation, aneuploidy, and persistence (Gow et al. 2022).

As fungal resistance has taken on alarming proportions and has grown substantially since 1960, the World Health Organization has called for the so-called triad solution. This strategy operates in three distinct but complementary areas: (1) The development of new antimicrobial agents. (2) The rational use of antimicrobial agents and informed prescribing. (3) Raising awareness among the target public (Rodrigues 2020).

When it comes to developing new antifungal agents, new delivery systems for antifungal agents have been studied, such as liposomes, carriers based on solid lipids and nanostructure lipids, polymeric nanoparticles and dendrimers (Voltan et al. 2016). These delivery strategies reduce the limitations that have been extensively described in this chapter, as well as reducing the serious systemic side effects of conventional drugs. However, they have a high added value when it comes to obtaining them.

With a view to the rational use of antimicrobial agents, conscious prescribing has been encouraged. In addition, with the implementation of the Global Action Plan on Antimicrobial Resistance, the aim has been to strengthen knowledge through surveillance and research. About raising awareness among the target public, information strategies have been used to improve understanding of antimicrobial resistance, reducing the incidence of infection. (https://www.who.int/news-room/fact-sheets/detail/antimicrobial-resistance, https://ahpsr.who.int/publications/i/item/global-action-plan-on-antimicrobial-resistance).

To summarize, the CNS is an important target for both fungal infections and antifungal agents. Therefore, to choose the most appropriate therapy, it is necessary to have an in-depth knowledge of the pathogen in question and the medicine to be used. It is highly important to consider the molecular structure (in particular its physico-chemical properties) in order to understand its ability to permeate the CNS, providing sufficient bioavailability to reach the therapeutic concentration, and finally, it is necessary to know the spectrum of action of the molecule on a given pathogen. In addition, the development of new medicines and advances in the metabolic engineering of microorganisms have generated promising prospects in the field of treating fungal infections that affect the CNS.

## References

Abbott NJ, Patabendige AAK, Dolman DEM, Yusof SR, Begley DJ (2010) Structure and function of the blood–brain barrier. Neurobiol Dis 37(1):13–25

Alegretti AP, Mucenic T, Brenol JCT, Xavier RM (2009) The role of CD55/CD59 complement regulatory proteins on peripheral blood cells of systemic lupus erythematosus patients. Rev Bras Reumatol [Internet] [cited 2023 Sep 2];49(3):276–287. Available from: https://www.scielo.br/j/rbr/a/wnCT8r7nCywSkY7ZSDbP65p/?lang=en

Arvanitis M, Anagnostou T, Fuchs BB, Caliendo AM, Mylonakis E (2014) Molecular and nonmolecular diagnostic methods for invasive fungal infections. Clin Microbiol Rev [Internet]. [cited 2023 Jun 23];27(3):490. Available from: /pmc/articles/PMC4135902/

Asiegbu FO, Kovalchuk A (2021) An introduction to forest biome and associated microorganisms. In: Forest microbiology: Volume 1: Tree microbiome: phyllosphere, endosphere and rhizosphere. Elsevier, pp 3–16

Barbosa PM, Barbosa ER (2020) The gut brain-axis in neurological diseases. Int J Cardiovasc Sci [Internet]. [cited 2023 Sep 2];33:528–536. Available from: https://www.scielo.br/j/ijcs/a/qkyyzLH3HS9QLbMvYN9V85N/

Bartemes KR, Kita H (2018) Innate and adaptive immune responses to fungi in the airway. J Allergy Clin Immunol [Internet]. [cited 2023 Sep 14];142(2):353. Available from: /pmc/articles/PMC6083885/

Bellmann R, Smuszkiewicz P (2017) Pharmacokinetics of antifungal drugs: practical implications for optimized treatment of patients. Infection 45:6. [Internet]. 2017 Jul 12 [cited 2023 Oct 13];45(6):737–79. Available from: https://link.springer.com/article/10.1007/s15010-017-1042-z

Benhar I, London A, Schwartz M (2012) The privileged immunity of immune privileged organs: the case of the eye. Front Immunol 3(Sep):296

Bongomin F, Gago S, Oladele RO, Denning DW (2017) Global and multi-national prevalence of fungal diseases—estimate precision. J Fungi 3:57. [Internet]. 2017 Oct 18 [cited 2023 Jun 23];3(4):57. Available from: https://www.mdpi.com/2309-608X/3/4/57/htm

Boynton TT, Ferneini EM (2016) Antimicrobial pharmacology for head, neck, and orofacial non-bacterial infections. In: Head, neck, and orofacial infections: an interdisciplinary approach, pp 164–173. https://doi.org/10.1016/B978-0-323-28945-0.00009-0

Brioschi S, Wang WL, Peng V, Wang M, Shchukina I, Greenberg ZJ et al (2021) Heterogeneity of meningeal B cells reveals a lymphopoietic niche at the CNS borders. Science [Internet]. [cited 2023 Sep 8];373(6553). Available from: /pmc/articles/PMC8448524/

Brumble LM, Reza MB, Dhakal LP, Cruz G, Abu Saleh OM, Heckman MG et al (2017) Fungal infections of the central nervous system: clinical, radiographic and laboratory manifestations. J Microbiol Exp [Internet] [cited 2023 Jun 23];5(6). Available from: https://medcraveonline.com/JMEN/JMEN-05-00167.php

Carabotti M, Scirocco A, Maselli MA, Severi C (2015) The gut-brain axis: interactions between enteric microbiota, central and enteric nervous systems. Ann Gastroenterol [Internet] [cited 2023 Sep 2];28(2):203. Available from: /pmc/articles/PMC4367209/

Carmona EM, Limper AH (2017) Overview of treatment approaches for fungal infections. Clin Chest Med 38(3):393–402

Casadevall A (2010) Cryptococci at the brain gate: break and enter or use a Trojan horse? J Clin Invest [Internet]. [cited 2023 Jun 23];120(5):1389. Available from: /pmc/articles/PMC2860912/

Casadevall A (2018) Fungal diseases in the 21st century: the near and far horizons. Pathog Immun [Internet] [cited 2023 Sep 9];3(2):183. Available from: /pmc/articles/PMC6241320/

Castellani G, Croese T, Peralta Ramos JM, Schwartz M (1979) Transforming the understanding of brain immunity. Science [Internet]. 2023 [cited 2023 Jul 23];380(6640). Available from: https://www.science.org/doi/10.1126/science.abo7649

Chi H, Pepper M, Thomas PG (2024) Principles and therapeutic applications of adaptive immunity. Cell [Internet] [cited 2024 May 25];187(9):2052–2078. Available from: http://www.cell.com/article/S0092867424003532/fulltext

Colombo AC, Rodrigues ML (2015) Fungal colonization of the brain: anatomopathological aspects of neurological cryptococcosis. An Acad Bras Cienc [Internet]. [cited 2023 Jun 23];87(2 Suppl):1293–1309. Available from: https://pubmed.ncbi.nlm.nih.gov/26247147/

Cornely OA, Arikan-Akdagli S, Dannaoui E, Groll AH, Lagrou K, Chakrabarti A et al (2014) ESCMID and ECMM joint clinical guidelines for the diagnosis and management of mucormycosis 2013. Clin Microbiol Infect [Internet]. [cited 2023 Jun 23];20 Suppl 3(S3):5–26. Available from: https://pubmed.ncbi.nlm.nih.gov/24479848/

Couves EC, Gardner S, Voisin TB, Bickel JK, Stansfeld PJ, Tate EW et al (2023) Structural basis for membrane attack complex inhibition by CD59. Nat Commun [Internet] [cited 2023 Sep 2];14(1). Available from: /pmc/articles/PMC9935631/

Cugurra A, Mamuladze T, Rustenhoven J, Dykstra T, Beroshvili G, Greenberg ZJ et al (2021) Skull and vertebral bone marrow are myeloid cell reservoirs for the meninges and CNS parenchyma. Science [Internet] [cited 2023 Sep 8];373(6553). Available from: /pmc/articles/PMC8863069/

Dallasta LM, Pisarov LA, Esplen JE, Werley JV, Moses AV, Nelson JA et al (1999) Blood-brain barrier tight junction disruption in human immunodeficiency Virus-1 encephalitis. Am J Pathol [Internet] [cited 2023 Sep 9];155(6):1915. Available from: /pmc/articles/PMC1866950/

Daneman R (2012) The blood-brain barrier in health and disease. Ann Neurol [Internet]. [cited 2023 Mar 27];72(5):648–672. Available from: https://pubmed.ncbi.nlm.nih.gov/23280789/

Daneman R, Prat A (2015a) The blood–brain barrier. Cold Spring Harb Perspect Biol [Internet]. [cited 2023 Mar 27];7(1). Available from: /pmc/articles/PMC4292164/

Daneman R, Prat A (2015b) The blood–brain barrier. Cold Spring Harb Perspect Biol 7(1):a020412

Drummond RA (2023) What fungal CNS infections can teach us about neuroimmunology and CNS-specific immunity. Semin Immunol 67:101751

Elmer BM, McAllister AK (2012) Major histocompatibility complex I proteins in brain development and plasticity. Trends Neurosci [Internet]. [cited 2023 Aug 27];35(11):660. Available from: /pmc/articles/PMC3493469/

Felton T, Troke PF, Hope WW (2014) Tissue penetration of antifungal agents. Clin Microbiol Rev [Internet] [cited 2023 Oct 13];27(1):68–88. Available from: https://journals.asm.org/doi/10.1128/cmr.00046-13

Fisher MC, Alastruey-Izquierdo A, Berman J, Bicanic T, Bignell EM, Bowyer P et al (2022) Tackling the emerging threat of antifungal resistance to human health. Nat Rev Microbiol 20:9. [Internet]. [cited 2023 Oct 13];20(9):557–571. Available from: https://www.nature.com/articles/s41579-022-00720-1

Forrester JV, McMenamin PG, Dando SJ (2018) CNS infection and immune privilege. Nat Rev Neurosci 19:11 [Internet]. [cited 2023 Apr 5];19(11):655–671. Available from: https://www.nature.com/articles/s41583-018-0070-8

Galea I, Bechmann I, Perry VH (2007) What is immune privilege (not)? Trends Immunol [Internet]. [cited 2023 Apr 5];28(1):12–18. Available from: http://www.cell.com/article/S1471490606003267/fulltext

Galgiani JN, Ampel NM, Blair JE, Catanzaro A, Geertsma F, Hoover SE et al (2016) Infectious Diseases Society of America (IDSA) Clinical practice guideline for the treatment of coccidioidomycosis. Clin Infect Dis [Internet]. [cited 2023 Jun 23];63(6):e112–e146. Available from: https://pubmed.ncbi.nlm.nih.gov/27470238/

Gavito-Higuera J, Mullins C, Ramos-Duran L, Olivas Chacon C, Hakim N, Palacios E (2016a) Fungal infections of the central nervous system: a pictorial review. J Clin Imaging Sci [Internet] [cited 2023 Sep 9];6(2). Available from: /pmc/articles/PMC4926551/

Gavito-Higuera J, Mullins C, Ramos-Duran L, Olivas Chacon C, Hakim N, Palacios E (2016b) Fungal infections of the central nervous system: a pictorial review. J Clin Imaging Sci [Internet]. [cited 2023 Jun 23];6(2). Available from: /pmc/articles/PMC4926551/

Ghannam JY, Al Kharazi KA (2023) Neuroanatomy, cranial meninges. StatPearls [Internet]. [cited 2023 Sep 7]. Available from: https://www.ncbi.nlm.nih.gov/books/NBK539882/

Góralska K, Blaszkowska J, Dzikowiec M (2018a) Neuroinfections caused by fungi. Infection [Internet]. [cited 2023 Jun 23];46(4):443. Available from: /pmc/articles/PMC6096918/

Góralska K, Blaszkowska J, Dzikowiec M (2018b) Neuroinfections caused by fungi. Infection 46:4. [Internet]. 2018 May 21 [cited 2023 Oct 13];46(4):443–459. Available from: https://link.springer.com/article/10.1007/s15010-018-1152-2

Gow NAR, Latge JP, Munro CA (2017) The fungal cell wall: structure, biosynthesis, and function. Microbiol Spectr [Internet] [cited 2023 Sep 9];5(3). Available from: https://journals.asm.org/doi/10.1128/microbiolspec.funk-0035-2016

Gow NAR, Johnson C, Berman J, Coste AT, Cuomo CA, Perlin DS et al (2022) The importance of antimicrobial resistance in medical mycology. Nat Commun 13:1. [Internet]. 2022 Sep 12 [cited 2023 Oct 13];13(1):1–12. Available from: https://www.nature.com/articles/s41467-022-32249-5

Guarner J, Brandt ME (2011) Histopathologic diagnosis of fungal infections in the 21st century. Clin Microbiol Rev [Internet]. [cited 2023 Jun 23];24(2):247–280. Available from: https://pubmed.ncbi.nlm.nih.gov/21482725/

Herisson F, Frodermann V, Courties G, Rohde D, Sun Y, Vandoorne K et al (2018) Direct vascular channels connect skull bone marrow and the brain surface enabling myeloid cell migration. Nat Neurosci [Internet]. [cited 2023 Sep 8];21(9):1209. Available from: /pmc/articles/PMC6148759/

Holling TM, Schooten E, Van Den Elsen PJ (2004) Function and regulation of MHC class II molecules in T-lymphocytes: of mice and men. Hum Immunol 65(4):282–290

Hong S, Van Kaer L (1999) Immune privilege: keeping an eye on natural killer T cells. J Exp Med [Internet]. [cited 2023 Apr 5];190(9):1197. Available from: /pmc/articles/PMC2195673/

Hubbard JA, Binder DK (2016) Inflammation. Astrocytes and Epilepsy [Internet] [cited 2023 Apr 5];313–342. Available from: https://linkinghub.elsevier.com/retrieve/pii/B9780128024010000132

Jiang S (2016) Immunity against fungal infections. Immunol Immunogenet Insights [Internet]. [cited 2023 Sep 14];8:3–6. Available from: https://journals.sagepub.com/doi/full/10.4137/III.S38707

Jong A, Wu CH, Prasadarao NV, Kwon-Chung KJ, Chang YC, Ouyang Y et al (2008) Invasion of C. neoformans into human brain microvascular endothelial cells requires protein kinase C-α activation. Cell Microbiol [Internet]. [cited 2023 Jun 23];10(9):1854. Available from: /pmc/articles/PMC2729555/

Kadry H, Noorani B, Cucullo L (2020) A blood–brain barrier overview on structure, function, impairment, and biomarkers of integrity. Fluids Barriers CNS [Internet]. [cited 2023 Sep 8];17(1):69. Available from: /pmc/articles/PMC7672931/

Köhler JR, Casadevall A, Perfect J (2015) The spectrum of fungi that infects humans. Cold Spring Harb Perspect Med [Internet]. [cited 2023 Jun 23];5(1). Available from: /pmc/articles/PMC4292074/

Konkel NL (2024) Wake-up call: rapid increase in human fungal diseases under climate change. Environ Health Perspect [Internet]. [cited 2024 May 29];132(4):42001. Available from: /pmc/articles/PMC11034633/

Koutsouras GW, Ramos RL, Martinez LR (2017) Role of microglia in fungal infections of the central nervous system. Virulence [Internet]. [cited 2023 Jun 23];8(6):705. Available from: /pmc/articles/PMC5626199/

Kristensson K (2011) Microbes' roadmap to neurons. Nat Rev Neurosci 12:345–357

Kwon D (2022) Guardians of the brain: how a special immune system protects our grey matter. Nature 606(7912):22–24

Liddelow SA (2015) Development of the choroid plexus and blood-CSF barrier. Front Neurosci 9(Mar):32

Liu Y, Feng M, Yao Y, Deng K, Bao X, Liu X et al (2017) Cryptococcal meningitis after transnasal transsphenoidal pituitary microsurgery of ACTH-secreting pituitary adenoma: a case report. Medicine [Internet] [cited 2023 Jun 23];96(28). Available from: /pmc/articles/PMC5515736/

London A, Benhar I, Schwartz M (2012) The retina as a window to the brain—from eye research to CNS disorders. Nat Rev Neurol 9:1. [Internet]. 2012 Nov 20 [cited 2023 Aug 27];9(1):44–53. Available from: https://www.nature.com/articles/nrneurol.2012.227

Massrey C, Altafulla JJ, Iwanaga J, Litvack Z, Ishak B, Oskouian RJ et al (2018) Variations of the transverse sinus: review with an unusual case report. Cureus [Internet]. [cited 2023 Sep 7];10(9). Available from: /pmc/articles/PMC6217863/

McCarthy M, Rosengart A, Schuetz AN, Kontoyiannis DP, Walsh TJ (2014) Mold infections of the central nervous system. N Engl J Med [Internet]. [cited 2023 Jun 23];371(2):150. Available from: /pmc/articles/PMC4840461/

Medawar PB (1948) Immunity to homologous grafted skin. III. The fate of skin homographs transplanted to the brain, to subcutaneous tissue, and to the anterior chamber of the eye. Br J Exp Pathol [Internet] [cited 2023 Aug 27];29(1):58. Available from: https://www.ncbi.nlm.nih.gov/pmc/articles/PMC2073079/

Merle NS, Church SE, Fremeaux-Bacchi V, Roumenina LT (2015) Complement system part I–molecular mechanisms of activation and regulation. Front Immunol 6(Jun):134383
Mohamed SH, Nyazika TK, Ssebambulidde K, Lionakis MS, Meya DB, Drummond RA (2022) Fungal CNS infections in Africa: the neuroimmunology of cryptococcal meningitis. Front Immunol [Internet]. [cited 2024 May 29];13. Available from: /pmc/articles/PMC9010970/
Mohamed SH, Fu MS, Hain S, Alselami A, Vanhoffelen E, Li Y et al (2023) Microglia are not protective against cryptococcal meningitis. Nat Commun 14:1. [Internet]. 2023 Nov 8 [cited 2024 May 29];14(1):1–15. Available from: https://www.nature.com/articles/s41467-023-43061-0
Mulvihill JJ, Cunnane EM, Ross AM, Duskey JT, Tosi G, Grabrucker AM (2020) Drug delivery across the blood-brain barrier: recent advances in the use of nanocarriers. Nanomedicine (Lond) [Internet]. [cited 2023 Oct 14];15(2):205–214. Available from: https://pubmed.ncbi.nlm.nih.gov/31916480/
Muñoz-Wolf N, Lavelle EC (2016) Innate immune receptors. Methods Mol Biol [Internet]. [cited 2023 Sep 13];1417:1–43. Available from: https://pubmed.ncbi.nlm.nih.gov/27221479/
Murthy JMK, Sundaram C (2014a) Fungal infections of the central nervous system. Handb Clin Neurol [Internet]. [cited 2023 Jun 23];121:1383–1401. Available from: https://pubmed.ncbi.nlm.nih.gov/24365427/
Murthy JMK, Sundaram C (2014b) Fungal infections of the central nervous system. Handb Clin Neurol 121:1383–1401
Naish E, Wood AJT, Stewart AP, Routledge M, Morris AC, Chilvers ER et al (2023) The formation and function of the neutrophil phagosome. Immunol Rev [Internet] [cited 2023 Sep 14];314(1):158–180. Available from: https://onlinelibrary.wiley.com/doi/full/10.1111/imr.13173
Niederkorn JY (2006) See no evil, hear no evil, do no evil: the lessons of immune privilege. Nat Immunol 7:4. [Internet]. 2006 [cited 2023 Apr 4];7(4):354–9. Available from: https://www.nature.com/articles/ni1328
O'Hanlon SJ, Rieux A, Farrer RA, Rosa GM, Waldman B, Bataille A et al (2018) Recent Asian origin of chytrid fungi causing global amphibian declines. Science (1979) [Internet]. [cited 2023 Sep 9];360(6389):621–627. Available from: https://www.science.org/doi/10.1126/science.aar1965
Palmano K, Rowan A, Guillermo R, Guan J, McJarrow P (2015) The role of gangliosides in neurodevelopment. Nutrients [Internet]. [cited 2023 Sep 2];7(5):3891. Available from: /pmc/articles/PMC4446785/
Pappas G, Vrioni G (2024) The last of us and the question of a fungal pandemic in real life. Emerg Infect Dis [Internet]. [cited 2024 May 29];30(3):595. Available from: /pmc/articles/PMC10902526/
Patin EC, Thompson A, Orr SJ (2019) Pattern recognition receptors in fungal immunity. Semin Cell Dev Biol [Internet]. [cited 2023 Sep 9];89:24. Available from: /pmc/articles/PMC6461132/
Patterson TF, Thompson GR, Denning DW, Fishman JA, Hadley S, Herbrecht R et al (2016) Practice guidelines for the diagnosis and management of aspergillosis: 2016 update by the Infectious Diseases Society of America. Clin Infect Dis [Internet] [cited 2023 Jun 23];63(4):e1–e60. Available from: https://pubmed.ncbi.nlm.nih.gov/27365388/
Rajshekhar V (2007) Surgical management of intracranial fungal masses. Neurol India [Internet] [cited 2023 Jun 23];55(3):267–273. Available from: https://pubmed.ncbi.nlm.nih.gov/17921656/
Raman SR (2010) Fungal infections of the nervous system: current perspective and controversies in management. Int J Surg 8(8):591–601
Robbins N, Wright GD, Cowen LE (2016) Antifungal drugs: the current armamentarium and development of new agents. Microbiol Spectr [Internet]. [cited 2023 Oct 13];4(5). Available from: https://journals.asm.org/doi/10.1128/microbiolspec.funk-0002-2016
Robbins N, Caplan T, Cowen LE (2017) Molecular evolution of antifungal drug resistance. Annu Rev Microbiol [Internet]. [cited 2023 Oct 13];71:753–775. Available from: https://pubmed.ncbi.nlm.nih.gov/28886681/

Rodrigues AG (2020) Yeast engineering for new antifungal compounds: a contextualized overview. [cited 2023 Oct 13];17–36. Available from: https://link.springer.com/chapter/10.1007/978-3-030-41870-0_2

Rueda R (2013) Gangliosides, immunity, infection and inflammation. In: Diet, immunity and inflammation. Woodhead Publishing, pp 341–358

Santiago-Tirado FH, Doering TL (2017) False friends: phagocytes as Trojan horses in microbial brain infections. PLoS Pathog [Internet] [cited 2023 Jun 23];13(12). Available from: /pmc/articles/PMC5730104/

Schwartz S, Kontoyiannis DP, Harrison T, Ruhnke M (2018) Advances in the diagnosis and treatment of fungal infections of the CNS. Lancet Neurol [Internet]. [cited 2023 Oct 13];17(4):362–372. Available from: http://www.thelancet.com/article/S1474442218300309/fulltext

Sharma RR, Pawar SJ, Lad SD, Mishra GP, Netalkar AS, Rege S (2012) Fungal infections of the central nervous system. In: Schmidek and sweet operative neurosurgical techniques: indications, methods, and results, vol 2, 6th edn, pp 1691–1732. https://doi.org/10.1016/C2011-1-05132-9

Shaz BH (2009) Other blood group systems, collections and antigens. In: Transfusion medicine and hemostasis: clinical and laboratory aspects. Academic Press, pp 145–149

Solaro C, Barratt D, Vaccarezza M (2022) The new anatomy of neuroimmunology. Immuno 2:255–259. [Internet]. [cited 2023 Sep 7];2(1):255–259. Available from: https://www.mdpi.com/2673-5601/2/1/16/htm

Speakman EA, Dambuza IM, Salazar F, Brown GD (2020) T cell antifungal immunity and the role of C-type lectin receptors. Trends Immunol [Internet]. [cited 2024 May 25];41(1):61. Available from: /pmc/articles/PMC7427322/

Stewart GG (2017a) History of brewing and distilling yeast. Brewing and distilling yeasts [Internet] [cited 2023 Sep 13];11–36. Available from: https://link.springer.com/chapter/10.1007/978-3-319-69126-8_2

Stewart GG (2017b) The structure and function of the yeast cell wall, plasma membrane and periplasm. Brewing and Distilling Yeasts [Internet]. [cited 2023 Sep 13];55–75. Available from: https://link.springer.com/chapter/10.1007/978-3-319-69126-8_5

Stockamp NW, Thompson GR (2016) Coccidioidomycosis. Infect Dis Clin N Am 30(1):229–246

Su Y, Zheng H, Shi C, Li X, Zhang S, Guo G et al (2023) Meningeal immunity and neurological diseases: new approaches, new insights. J Neuroinflammation [Internet] [cited 2023 Sep 8];20(1):1–12. Available from: https://jneuroinflammation.biomedcentral.com/articles/10.1186/s12974-023-02803-z

Sun L, Su Y, Jiao A, Wang X, Zhang B (2023) T cells in health and disease. Signal Transduction and Targeted Therapy 2023 8:1 [Internet]. [cited 2024 May 25];8(1):1–50. Available from: https://www.nature.com/articles/s41392-023-01471-y

Talbot S, Foster SL, Woolf CJ (2016) Neuroimmunity: physiology and pathology. Annu Rev Immunol [Internet]. [cited 2024 May 28];34:421–447. Available from: https://www.annualreviews.org/content/journals/10.1146/annurev-immunol-041015-055340

Taylor AW, Kaplan HJ (2010) Ocular immune privilege in the year 2010: ocular immune privilege and uveitis. Ocul Immunol Inflamm [Internet]. [cited 2023 Aug 27];18(6):488. Available from: /pmc/articles/PMC3329254/

Upadhyay RK (2014) Drug delivery systems, CNS protection, and the blood brain barrier. Biomed Res Int [Internet]. [cited 2023 Mar 27];2014. Available from: /pmc/articles/PMC4127280/

Voltan AR, Quindós G, Alarcón KPM, Fusco-Almeida AM, Mendes-Giannini MJS, Chorilli M (2016) Fungal diseases: could nanostructured drug delivery systems be a novel paradigm for therapy? Int J Nanomedicine [Internet]. [cited 2023 Oct 13];11:3715–3730. Available from: https://www.tandfonline.com/action/journalInformation?journalCode=dijn20

Weishaupt N, Zhang A (2016) Why doesn't your brain heal like your skin? Front Young Minds 4:a020602

Wieczorek M, Abualrous ET, Sticht J, Álvaro-Benito M, Stolzenberg S, Noé F et al (2017) Major histocompatibility complex (MHC) class I and MHC class II proteins: conformational plasticity in antigen presentation. Front Immunol 8(Mar):292

Williams PB, Barnes CS, Portnoy JM, Baxi S, Grimes C, Horner WE et al (2016) Innate and adaptive immune response to fungal products and allergens. J Allergy Clin Immunol Pract 4(3):386–395

Wirth F, Ishida K (2020) Antifungal drugs: an updated review of central nervous system pharmacokinetics. Mycoses [Internet] [cited 2023 Oct 13];63(10):1047–1059. Available from: https://onlinelibrary.wiley.com/doi/full/10.1111/myc.13157

Woodring T, Deepe GS, Levitz SM, Wuethrich M, Klein BS (2023a) They shall not grow mold: soldiers of innate and adaptive immunity to fungi. Semin Immunol [Internet]. [cited 2023 Sep 9];65:101673. Available from: /pmc/articles/PMC10311222/

Woodring T, Deepe GS, Levitz SM, Wuethrich M, Klein BS (2023b) They shall not grow mold: soldiers of innate and adaptive immunity to fungi. Semin Immunol 65:101673

Wüthrich M, Deepe GS, Klein B (2012) Adaptive immunity to fungi. Annu Rev Immunol [Internet] [cited 2023 Sep 14];30:115. Available from: /pmc/articles/PMC3584681/

Young TL, Zychowski KE, Denson JL, Campen MJ (2019) Blood-brain barrier at the interface of air pollution-associated neurotoxicity and neuroinflammation. Adv Neurotoxicol 3:295–337

Zlokovic BV (2008) The blood-brain barrier in health and chronic neurodegenerative disorders. Neuron [Internet] [cited 2023 Mar 27];57(2):178–201. Available from: https://pubmed.ncbi.nlm.nih.gov/18215617/

# Role of Vaccines and Monoclonal Antibodies in Systemic Candidiasis: Past and Future Approaches

**Pankaj Chandley and Soma Rohatgi**

## Introduction

### *Systemic Candidiasis*

The term "candidiasis" generically refers to fungal infections caused by fungi belonging to the genus *Candida*. *Candida* species have emerged as significant opportunistic infection-causing pathogens and can live within human hosts as commensals. When an individual's environment changes or their body's microbial flora overgrows or is disrupted, it usually results in superficial infections. However, in the event of tissue barrier breakdown or under immune-compromising circumstances, superficial *Candida* infections can cause the spread of fungus into blood, which is known as invasive or systemic candidiasis (François et al. 2013). It is among the most prevalent blood stream infections in hospitalized patients worldwide (Warnock 2007). Even with the usage of antifungal medications, the death rate due to systemic candidiasis range from 40% to 70% globally (Calandra et al. 2016). It is the fourth most prevalent nosocomial blood stream infection experienced by patients in intensive care units globally (Pappas et al. 2018). Over 250,000 people worldwide are affected by systemic candidiasis each year, which results in about 50,000 fatalities (Kullberg and Arendrup 2015). Five distinct species of *Candida*, including *C. albicans, C. glabrata, C. parapsilosis, C. tropicalis,* and *C. krusei,* are responsible for more than 90% of invasive infections. More recently, *C. auris*, a multi-medicine resistance species, has been connected to significant episodes of severe infections in hospitals worldwide (Ahmad and Alfouzan 2021). Presently, azoles, polyenes, echinocandins, allylamines, and antimetabolites are the five kinds of antifungal

P. Chandley · S. Rohatgi (✉)
Department of Biosciences and Bioengineering, Indian Institute of Technology Roorkee, Roorkee, Uttarakhand, India
e-mail: soma.rohatgi@bt.iitr.ac.in



A. Gomes Rodrigues, A. Gupta (eds.), *Fungal Infections*, Fungal Biology,
https://doi.org/10.1007/978-3-032-06014-3_8

medications used to treat invasive candidiasis (Gintjee et al. 2020). Despite advancements in antifungal medication, invasive candidiasis patients continue to have a remarkably high morbidity and fatality rate. Antifungal medication toxicity and negative side effects further restrict their utilization. Additionally, over the past 10 years, more *Candida* species have developed drug resistance to antifungal medications, and these species now pose a severe danger to human health on a global scale. Over 34,000 cases and 1700 fatalities each year were attributed to drug-resistant *Candida* species, as per CDC's data based on the issue of antibiotic resistance in year 2017 (Bhattacharya et al. 2020). A major source of worry is also the extensive spread and emergence of the novel *Candida* strains. In addition to antifungal medication resistance, long-term research has shown that over the past few decades, a shift toward non-albicans *Candida* (NAC) species has occurred (Papon et al. 2013). Population-based research indicates that there are regional differences in the geographic distribution of the NAC species and *C. albicans* (Whibley and Gaffen 2015).

There is a need for development of novel vaccinations against systemic *Candida* infection for patients at risk, such as premature neonates, cancer patients, immunocompromised patients, and those undergoing invasive therapies for extended durations in hospitals. Moreover, there is a critical requirement for the development of vaccines or immunotherapies that specifically target *Candida* species to alleviate the load and difficulties caused by *Candida*-related systemic candidiasis. Therefore, the treatment of systemic candidiasis brought on by *Candida* species requires new alternative immunotherapeutic methods. For a very long time, the function of humoral immunity in the host's defense against systemic candidiasis has been overshadowed by the function of cellular immunity. This chapter's primary goal is to share knowledge in public interest on the function of vaccines and protective antibodies targeting *Candida* vaccine candidates from experimental studies focusing primarily on *C. albicans*.

## Innate Immunity Against Systemic Candidiasis

The initial defense against fungi is considered physical barriers, such as skin and mucosal epithelial surfaces on the gastrointestinal, genitourinary, upper airway, and oral tracts (Nami et al. 2019). Both renal epithelial as well as endothelial cells of blood vessels have been linked to imparting protection against systemic candidiasis, however, epithelial cells have also been shown to play significant roles during anti-*Candida* responses (Lionakis 2014). The innate immune response of host not only offers a crucial early defense against fungi, but it also triggers several immune responses that are thereafter carried out by the adaptive immune system (Medici and Del Poeta 2015). As the initial step in innate immunity activation, pathogen-associated molecular patterns (PAMPs) expressed on *Candida* are recognized by pattern recognition receptors, which include lectins, integrins, and toll-like receptors. These pattern recognition receptors act as effective receptors for the phagocytosis of opsonized fungus and upregulation of proinflammatory cytokines for facilitating phagocyte migration to the site of infection (Vonk et al. 2006). Various investigations utilizing complement protein C3 depletion in animal models have identified a role of the complement system in host defense against *Candida* (Harpf et al. 2020; Tsoni et al. 2009). Anaphylatoxins (C3a, C4a, and C5a) are produced when the complement system is activated along with their ability to promote phagocytosis and play a role in many different modes of antifungal action (Brown et al. 2012). The function of phagocytic cells, including neutrophils, macrophages, monocytes, and dendritic cells (DCs), in the body's innate defense against systemic candidiasis has been well established. Macrophages shows the ability to phagocytose and destroy *Candida* cells, reducing the fungal burden soon after infection (Jones-Carson et al. 1997). Early in an infection, when *Candida* is present, monocytes are more successful in killing it than DCs or macrophages are (Netea et al. 2004). Mice deficient in monocytes are more likely to contract *C. albicans* infections, and *C. albicans* exposure causes monocytes to release tumor necrosis factor (TNF-$\alpha$), which is necessary for preventing systemic *Candida* infection (Netea et al. 1999; Ngo et al. 2014). Additionally, human blood monocytes have been shown to have candidacidal action (Netea et al. 2015). Due to their capacity for phagocytosis and antigen presentation, DCs contribute significantly to the host's defense against systemic candidiasis (Newman and Holly 2001; Romagnoli et al. 2004). Following phagocytosis, DCs produce antagonistic types of cytokines (Th1/Th2) that allow them to distinguish between yeast and hyphae. Although *Candida* can be ingested and killed by DCs, macrophages have been shown to be more effective at killing fungi (Netea et al. 2004). Moreover, neutrophils play a significant role in mediating immunity against *Candida* infections and their activation is necessary for the *Candida* removal. On the other hand, neutropenia is a significant risk factor contributing to invasive fungal infections (Gardner (2007); Netea et al. 2015). The ability of neutrophils to successfully prevent the transition of *Candida* blastopores into hyphae, a crucial aspect of the fungus' pathogenicity, has been shown as a function of their role in facilitating *Candida* killing (Lionakis and Netea 2013). A key part of the initial response against mouse systemic *C. albicans* infection is played by natural killer (NK) cells. T/B/NK-cell defective mice were found to be susceptible to primary and secondary systemic candidiasis than T or B-cell deficient SCID mice, according to research (Quintin et al. 2014). Human NK cells have been used to evaluate the impact of NK cells in systemic candidiasis; they may exert direct cytotoxic activity on the fungi (Voigt et al. 2018).

## *Adaptive Immunity Against Systemic Candidiasis*

### Cellular Responses Against Systemic Candidiasis

Adaptive immunity, in addition to innate immunity, is crucial for the efficient eradication of the best preventive immunity to invasive candidiasis. According to research, the most crucial host's defense systems against systemic candidiasis is cell-mediated adaptive immunological responses, in which $CD4^+$ and $CD8^+$ T cells are the key mediators of protection against candidiasis. In the past, it was thought that $CD4^+$ T cell immune responses largely offered protection from *Candida* infection. During an antifungal immune response, $CD4^+$ T cells are widely attributed to Th1, Th2, and Th17-mediated immune responses (Conti and Gaffen 2010). The direct cytotoxic activity and cytokine release of $CD8^+$ T cells, on the other hand, is shown to confer immunity against systemic fungal infections (Ashman et al. 1999). The essential mechanism of $CD8^+$ T cells in direct killing of *Candida* has been well documented. Also, various cytokine secretions such as TNF-α and interferon gamma, (IFN-γ) have been demonstrated as major effector mechanisms through which cytotoxic $CD8^+$ T cells can prevent *Candida* infection (Daniel Gozalbo 2009). There are numerous cytokines and chemokines linked to defense against *Candida* infection (Shukla et al. 2021). According to studies, some cytokines help phagocytes destroy *Candida* species, and others are essential for T-helper (Th) cell development in antigen-presenting cells including DCs and macrophages (Hamad 2012). The cytokines that Th1 cells secrete can also trigger B cells, causing them to secrete antibodies against *Candida* that are specific to the antigen. Additionally, Th1/Th17-mediated immunity has been strongly associated with resistance to fungi infections (van de Veerdonk et al. 2012).

## Humoral Responses Against Systemic Candidiasis

Although cellular immunity is known to be the major element in systemic candidiasis, humoral immunity also plays a key role in preventing invasive *Candida* infections. Although humoral immunity includes the collectins, complement system, and antimicrobial peptides, this section's main objective is to assess the humoral responses generated by B cells and antibodies that provide resistance during systemic *Candida* infection.

### B Cell Responses in Systemic Candidiasis

An earlier investigation revealed that mice with B-cell depletion were no longer susceptible to *Candida* infection (Carrow et al. 1984). It was discovered in a previously published study that B-cell depletion did not make the mice more susceptible to *C. albicans*-mediated candidiasis (Sinha et al. 1987). A study also revealed that B lymphocytes were not able to play a key role in the defense against the *C. albicans* (Bistoni et al. 1988). It was reported that B-cell depletion did not enhance propensity to *C. albicans* infection, according to previously reported investigations using SCID mice (Balish et al. 1993). However, a study utilizing germ-free B cell deletion mice revealed that these animals are immune to mucosal candidiasis but sensitive to experimental systemic candidiasis (Wagner et al. 1996). To increase defenses against systemic candidiasis during B cell deficiency, thymic and extrathymic T cells participated in host defense to *C. albicans* (Jones-Carson et al. 1997). It has been observed that animals obtaining vaginal $CD3^{-}CD5^{+}$ B cells from immunized rats had less *Candida* colony forming units than controls, although they displayed a significantly slower rate of fungal clearance than animals receiving immune T cells (Santoni et al. 2002). Another investigation showed that vaginal B cells from rats that had been exposed to *Candida* led to protection against vaginal candidiasis in naive animals (De Bernardis et al. 2010). Additionally, rituximab was added in vitro to PBMCs (peripheral blood mononuclear cells) for the depletion of B cells, which decreased the amount of anti-*Candida* responses (Van De Veerdonk et al. 2011). Adaptive immunity was not necessary for mice lacking T and B cells (Rag-1 knockouts) to survive both the early *C. dubliniensis* or *S. aureus* challenge and the subsequent challenge (Lilly et al. 2018). Previous research has linked B1 B-cells and naturally occurring IgM generated from B1 to anti-fungal immunity (Li et al. 2007; Rapaka et al. 2010). According to a study, TgVH3B4 animals eliminated *C. albicans* more quickly than control mice after receiving a fungal inoculation. Notably, *C. albicans* infection increased B1 cell proliferation, resulting in a rise in B1a B-cells and *C. albicans*-specific B cells in TgVH3B4 mice, which could aid in the clearance of fungi (Tian et al. 2013). According to a recent study, antibodies created by cloning immunoglobulin genes from B cell cultures made from *C. albicans*-infected patients were able to both incite opsonophagocytic activity of macrophage in vitro and protect against invasive candidiasis under in vivo setting (Rudkin et al. 2018). Also, human B cells have been found to have an antibody-independent role involving cytokine production, which can support antifungal immunity (Li et al. 2017).

## Antibody Responses in Systemic Candidiasis

The key players of antifungal adaptive immunity, that limit fungal load and assist in its clearance, are antibodies. Among other effector immune responses, antibodies can neutralize toxins, inhibit pathogens from adhering to host cells, opsonize, and cause antibody dependent cellular cytotoxicity (ADCC), (Casadevall et al. 2002). Further anti-fungal antibody immune responses which are specific to *Candida* are inhibition of biofilm development, inhibition of germ tube formation, phagocytosis, complement activation, immunological modulation, and direct fungal growth inhibition (Shukla et al. 2021).

The primary antibodies generated against fungi are of the isotypes IgM, IgG, and IgA. Among these, IgA was the major antibody implicated in imparting immunity to mucosal surfaces. It is also known to inhibit *Candida* attachment to human oral epithelial cells (Vudhichamnong et al. 1982). According to Maiti et al., controls with functional B-cells produced an effective antibody response against systemic candidiasis, but animals lacking B-cells were unable to produce such immune response. The body weight, mortality risk, and susceptibility to *Candida* infection were all higher in the B-cell deficient mice. This finding revealed that the defense against *Candida* infection is provided by B-cells and antibodies (Maiti et al. 1985). Studies have been done on the function of anti-*Candida* antibodies in the passive administration and defense of the host against systemic candidiasis (Cassone et al. 2005). But it is still unknown how vaginal antibodies contribute to the development of vaginal candidiasis (Cassone et al. 2007). According to one study, the levels of anti-*Candida* IgA and IgG antibodies were identical in patients with and without vaginal candidiasis. Additionally, the anti-*Candida* IgA antibody may not have been protective against repeated vaginal *Candida* infection (Bohler et al. 1994). Another study characterized the levels of anti-*Candida* IgE, and IgA, IgG antibodies in sera and vaginal wash among women with and without vaginal candidiasis. It was found that patients had greater amounts of IgA in vaginal wash and reduced levels of IgA in serum. Moreover, patients with and without a past of *Candida* vaginitis showed comparable levels of blood anti-*Candida* IgA and IgG antibodies (de Carvalho et al. 2003).

According to a study, anti-*Candida* antibodies were able to prevent the fungus from adhering to the vaginal epithelium, which may have contributed to protecting rats from experimental vaginitis (De Bernardis et al. 2007). Earlier it had been shown that mice were protected from widespread candidiasis by antibodies made against specific cell surface antigens of *C. albicans* (Han and Cutler 1995). The host becomes protected from fungal infection by antibodies against fungus cell wall antigens such as various glycoproteins, and enzymes (Lopez-Ribot Jose et al. 2004). Also, cell wall surface proteins were included in the liposomal antigen delivery technique used to immunize mice, and these mice showed greater serum levels of *C. albicans*-specific antibodies (Carneiro et al. 2016). Additionally, vaccinations based on antibodies have been demonstrated to provide protection against systemic candidiasis (Xin and Cutler 2011).

## Monoclonal Antibodies Against Systemic Candidiasis

In the next few years, sales of monoclonal antibodies are expected to surpass 125 billion dollars globally, making them among of the most popular medicines in the world. Many of these monoclonal antibodies have currently been granted approval for the treatment of autoimmune disorders and cancer. Additionally, developments in the production of monoclonal antibodies to bacterial, and viral targets, as well as the exploration of antibody-antibiotic conjugates as potential therapies against intracellular bacterial infections, have also occurred in recent years.

However, mycotic infections have not yet been the focus of the breakthrough in clinical monoclonal antibody research. It has now been reported that protective monoclonal antibodies for clinically significant fungi exist, however, they are entirely of murine origin and produced using hybridoma technique. In recent years, there has been a significant diversification in the techniques and strategies used to produce monoclonal antibodies for diagnostic and/or therapeutic application. Although early monoclonal antibodies were of mouse origin, the human host was immunogenic.

Most monoclonal antibodies utilized in clinical settings today are IgG1 chimeric, humanized, or human antibodies produced using hybridoma cell lines. Fully human monoclonal antibodies have also been produced utilizing combinatorial technology using phage or yeast, although they frequently rely for an interval of in vitro affinity maturation and result in monoclonal antibodies with randomized heavy and light chain pairings. Complete human antibodies would be extremely important tools for investigating potential immunotherapies that target medical mycoses in the future (Rudkin et al. 2018). There are currently no FDA-approved therapeutic monoclonal antibodies for systemic candidiasis, despite extensive study into the development of monoclonal antibodies in preclinical investigations. There is an urgent need for developing monoclonal antibodies for the treatment of immunocompromised individuals. Therefore, this chapter outlines current developments in monoclonal antibodies against systemic candidiasis.

## Vaccine Candidates in Systemic Candidiasis

The need for anti-*Candida* vaccines has increased due to the growing incidence and associated mortality rates of systemic candidiasis, particularly in high-risk populations like the elderly, immunocompromised individuals, such as cancer patients, newborns, and individuals receiving invasive therapy for long duration in hospitals.

Even though multiple potential vaccine candidates against *Candida* have been investigated, only a small number of them have so far advanced to the level of clinical trials (Wan Tso et al. 2018). Numerous studies using animal models, has assessed the immunogenicity and effectiveness of various vaccine candidates against candidiasis. The clinical advancement of anti-*Candida* vaccines that effectively confer protective immunity to people at risk of contracting systemic fungal infections is still hindered by several obstacles (Wang et al. 2015). An efficient fungal vaccine must be able to elicit robust immune responses and immunological memory, that may also protect against recurring fungal infections. In this regard, live attenuated *Candida* strains, recombinant proteins, cell-wall polysaccharides, and glycoconjugates have all been examined as potential anti-*Candida* vaccines candidates. Additionally, a wide range of techniques, including adjuvants and administration methods, have been studied to increase the effectiveness of the vaccines. Although numerous *C. albicans* virulence antigens have been identified as potent vaccine candidates, there have been few investigations to assess the role of humoral immune responses in the protective immunological potential provided by them against experimental systemic candidiasis. Moreover, the protective function of humoral immune responses to systemic candidiasis is barely understood.

## *Mannan*

**Vaccine Research**

The primary components of polysaccharides, which make up around 80% of the fungal cell wall, are glucan, mannan, and chitin (Ishibashi et al. 2005). The most usual form of mannan is long N-linked polymers with several hundred mannose residues connected to mannoprotein species with high molecular weights.

Mannan is regarded as a key polysaccharide antigen for humoral immune responses since it is found on the cell wall of the yeast *C. albicans*. It is critical to identify soluble mannan in serum when diagnosing invasive candidiasis (Mikulska et al. 2010). One study showed that without the involvement of direct T/B cell interactions, the mannan antigen can directly activate human B-cells to form anti-mannan antibodies, while type 2 T-cell independent anti-mannan antibody synthesis requires cytokines released by T cells (Mangeney et al. 1989). Han et al. also found that in a mouse model of vaginal candidiasis, either the antibodies produced against the mannan vaccination or the delivery of anti-mannan antibodies might confer protection against vaginal *Candida* infection (Han et al. 1998). Zhang et al. discovered that complement activation against disseminated candidiasis is mediated by anti-mannan IgG antibodies' ability to start the classical route by C3 deposition on *C. albicans* (Zhang and Kozel 1997). Additionally, they found that anti-mannan IgG antibodies can activate the complement system's conventional and alternative pathways (Zhang et al. 1998).

In a different study, neutropenic rabbits were immunized by a synthetic epitope of mannan trisaccharide along with tetanus toxoid, which generated protective antibodies and aided in preventing *C. albicans* infection (Lipinski et al. 2012). IgG antibodies were discovered to relate to protective antifungal immunity, and antisera produced by vaccinating rabbits with a vaccine conjugate containing mannan and human serum albumin which was able to limit the growth of various *Candida* species (Machová et al. 2015). To identify the epitopes of monoclonal antibodies that are specific to the mannan in *C. albicans*, Sendid et al. utilized synthesized oligomannosides. Also, unique specificities of protective epitopes of 1,2 mannotriose were discovered by examining the anti-mannan antibody response in the serum of human patients with systemic candidiasis (Sendid et al. 2021). Preclinical research has examined a variety of mannan-based vaccine formulations, including mannan extracts, mannan-BSA conjugates, mannan-HSA conjugates, and mannan oligosaccharide conjugates (Wu et al. 2007, 2008; Paulovičová et al. 2007, 2010; Xin et al. 2008, 2012; Lipinski et al. 2011, 2012). It has been extensively reviewed that B cells and antibodies specific to mannan antigen immensely contribute to humoral immunity against systemic Candidiasis (Shukla et al. 2021).

**Monoclonal Antibodies**

In one study, mannan fractions were encapsulated in liposomes and given to mice as a vaccine, it was discovered that antibodies specific for the mannan fractions can be linked to enhanced resistance to systemic candidiasis. Anti-mannan antibodies, both polyclonal as well as monoclonal, have been found to impart protection against systemic candidiasis (Han and Cutler 1995). The anti-mannan IgM monoclonal antibodies namely, B6 and B6.1, underwent thorough analysis, and it was discovered that the mannan-(1-2)-linked mannotriose epitope produces protective antibodies (Han et al. 1997). The B6.1 antibody along with intact serum complement was found to provide protection by helping neutrophils in engulfing and eliminating yeast cells (Caesar-TonThat and Cutler 1997). Anti-mannan antibodies were able to transmit anti-*Candida* protection to uninfected, nonimmunized rats.

Moreover, these antibodies helped in acquiring protection against *Candida* infection in rat vaginitis model (Cassone et al. 1995). In a different investigation, it was discovered that the protective anti-mannan IgM monoclonal antibody (AF1) response was dependent on T-cells (De Bernardis et al. 1997). Anti-mannan antibodies were discovered to have protective potential that was reliant on serum titer, epitope specificity, and the ability to fix complement to the pathogen surface, facilitating increased phagocytosis and fungus killing. Mice vaccinated with the liposome-mannan vaccine produced the monoclonal antibody C3.1 (IgG3), which was then evaluated (Han et al. 2000, 2001; Cutler 2005). According to another study, anti-mannan antibodies seen in the serum of healthy individuals showed biological effects such as stimulating *C. albicans* opsonophagocytic killing and activating complement (Kozel et al. 2004). According to Zhang et al., the mouse complement system cascade must be activated for an anti-mannan monoclonal antibody (M1g1) to increase the ability of peritoneal macrophages to phagocytose *C. albicans*. This antibody protects the host against systemic candidiasis (Zhang et al. 2006). In a second study, it was found that M1g1 demonstrated a distinct Fc-independent effector function via regulating C3 deposition to *C. albicans* (Boxx et al. 2009). Another study reported that the IgG subclass of the M1g1 plays a crucial part in mediating resistance against systemic candidiasis (Nishiya et al. 2016).

## *β*-Glucan

### Vaccine Research

On *Candida* cell surface, there are structurally intricate homopolymers of glucose known as β-1,3-glucans that serve as the pathogen-associated molecular patterns. β-glucans are a good option for vaccination to prevent systemic candidiasis since they are the main extracellular polysaccharide antigen and play a substantial role in the activation of host-protective immune responses. Even though β-glucan is not very immunogenic, it can be coupled with substances like diphtheria toxoid to create a vaccine that is efficacious against candidiasis. For studying antifungal immune responses, model ligands for β-glucan include curdlan, zymosan, and pustulan. Using a mouse model of disseminated candidiasis, one study demonstrated that animals inoculated with *C. albicans* cells had a protective immune response by generating anti-β-glucan antibodies (Bromuro et al. 2002). Anti-β-(1,3)-glucan IgG antibodies conferred protection to mice exposed to *C. albicans* while anti-β-(1,6)-glucan antibodies were reported as non-protective, when curdlan and/or synthetic β-glucan oligosaccharides were used as vaccines (Bromuro et al. 2010).

Naturally animals and human serum contain anti-β-glucan antibodies, which, when stimulated by β-glucan-based vaccinations, can have fungicidal protective effects. Studies have revealed that anti- β-glucan-like peptide mimotopes and polyclonal, monoclonal, and recombinant antibodies can kill *C. albicans* in vivo and/or in vitro (Magliani et al. 2008). Additionally, it was shown that diverse animal serum samples containing β-glucan antibodies with strong affinity to solubilized *C. albicans* β-glucan were observed to be involved in the immune response to pathogenic fungus (Ishibashi et al. 2010).When anti-glucan antibody profiles of patients with and without candidemia were compared, it was discovered that candidemia patients had significant levels of antibodies to β-(1,6)-glucan and low levels of antibodies to β-(1,3)-glucan. Moreover, mannoprotein MP65 antibodies were found to significantly correlate with survival (Torosantucci et al. 2017). According to Hoogeboom et al., human B cells with heavy chain BCRs generated by the IGHV3-7 gene family and light chain BCRs by the IGKV2-24 gene family showed strong sensitivity for β-(1,6)-glucan, a key antigenic component of yeast species and filamentous fungi (Hoogeboom et al. 2013). Preclinical research has examined several vaccine formulations, including β-glucan-conjugate vaccine, linear β-(1-3)-nonaglucoside plus BSA (G9), β-glucan oligosaccharides with keyhole limpet hemocyanin (KLH), and OVA plus Curdlan. These findings demonstrated an impressive immune response driven by B cells and antibodies against systemic *Candida* infection (Shukla et al. 2021).

**Monoclonal Antibodies**

In a recent study, an anti-1,3-β-D-glucan mAb was fused with an anti-mannoprotein mAb recognized by MP65 and has been employed efficiently for identifying invasive *Candida* infections (Zito et al. 2016). The in vitro interactions between *Candida* and the mouse monoclonal antibodies (3G11 and 5H5) made against synthetic nona-β-(1,3)-D-glucoside coupled with BSA showed synergy with the antifungal fluconazole. Additionally, they demonstrated protective effect in vivo, pointing to their potential application in combination antifungal therapy (Matveev et al. 2019). Furthermore, one study discovered that a single-chain anti-idiotypic antibody against yeast-killing toxin, suppressed the activity of β-1,3-glucan synthase, having an inhibitory effect on the development of *Candida* (Selvakumar et al. 2006). Additional research uncovered peptides in the anti-idiotypic antibody that may show interaction with the β-glucan of *C. albicans* and impede the development of fungi (Kabir et al. 2011).

## *Laminarin*

**Vaccine Research**

Laminarin is a β-glucan substance that was extracted from the brown alga *Laminaria digitata* and consists of erratic β-(1,6) branches and β-(1,3) repeating units. It is a type of β-glucan, obtained from non-fungal sources and numerous studies have noted the protective effects of laminarin against invasive candidiasis. An anti-idiotypic antibody peptide fragment's therapeutic efficacy was originally documented by Polonelli et al. during an experimental case of systemic candidiasis. Strong candidacidal activity was exhibited by the peptide in vitro, and laminarin was able to neutralize it, indicating that the interactions among killer peptides and the β-glucan moiety located on *Candida* cell wall were what caused the candidacidal activity (Polonelli et al. 2003). In one experiment, mice were vaccinated with CRM197, a laminarin and diphtheria toxoid glycoconjugate vaccine that provided immunity against systemic candidiasis. Additionally, it was discovered that the laminarin-CRM197 conjugate vaccine, given to mice using the MF59 adjuvant, provided protective immunity against vaginal candidiasis, which was linked to the generation of anti-β-glucan IgG antibodies (Pietrella et al. 2010). Other laminarin-based vaccine formulations, including tetanus toxoid combined to laminarin tricomponent conjugate vaccine and LAM-CRT conjugate, have also been reported to induce protective B cell-mediated immune responses against systemic candidiasis (Shukla et al. 2021). Furthermore, it has been noted that immunocompetent people have lower levels of anti-laminarin antibodies in comparison to antibodies to mannan and β-glucan. Because the antifungal effectiveness of the glyco-conjugate vaccine was established by the development of anti-laminarin antibodies, effective antifungal immunization strategies in humans should aim to shift the equilibrium of anti-*Candida* antibodies towards anti-laminarin antibodies (Chiani et al. 2009).

**Monoclonal Antibodies**

A group of researchers developed a glyco-conjugate vaccine by fusing laminarin with CRM197. Mice immunization with glycoconjugate vaccine generated a protective antibody-mediated response against systemic *Candida* infection (Torosantucci et al. 2005). To provide passive defense against systemic candidiasis infection, a laminarin-binding monoclonal antibody, 2G8 (IgG2b) was reported. Mice that had never received immune serum (IgG fraction) before were protected (Torosantucci et al. 2009). Additional studies on the monoclonal antibodies 2G8 and IE12 reported that protection from anti-β-glucan antibodies was connected to binding specificity to 1,3-β-glucan epitopes and inhibition of *Candida* adhesion and proliferation, and that the isotype of anti-glucan antibodies may adversely impact the specificity of the β-glucan epitopes recognized by these antibodies (Torosantucci et al. 2009; Cassone et al. 1930). Recently, anti β-glucan 2G8 mAb-derived chimeric human-murine monoclonal antibodies were generated in plants and discovered to be protective in mucosal and systemic *Candida* infection (Capodicasa et al. 2011).

## *Hsp90*

**Vaccine Research**

Heat shock proteins (Hsps) interact with a wide range of varied regulators of cell signaling pathways to control fundamental physiological functions or pathogenicity. In *Candida* species, they are expressed as a response to cell cycle control, morphogenesis, drug resistance and apoptosis. Humans can be shielded from systemic candidiasis by Hsps, which the immune system frequently targets in systemic fungal infections. Hsp90 offers defense against *C. albicans* and acts as a connection between innate and acquired humoral immune responses. The recovery from a systemic infection caused by *C. albicans* was shown by Matthews et al. to produce antibodies that are specific for both *Candida* cells and Hsp90 (Matthews and Burnie 1992). Additionally, mice immunized with serum carrying Hsp90-specific antibodies outlived mice receiving regular human serum during systemic candidiasis (Matthews and Burnie 1992). Invasive candidiasis was treated with a mouse monoclonal antibody to a ubiquitous Hsp90 epitope (LKVIRK), which showed anti-*C. albicans* effects (Matthews et al. 1991a, b). In acute and chronic models of murine invasive candidiasis, an antibody against the Hsp90 epitope was tested, which showed significant clearance of *C. albicans* from renal tissue and an increase in mouse survival rates indicated that antibodies to Hsp90 epitopes could be protective in mouse model of invasive candidiasis (Matthews et al. 1991a, b, 1995). It has been demonstrated that Hsp90-based vaccination formulations, including recombinant Hsp90 protein, hybrid phage particles exhibiting the Hsp90 epitope, proteoliposomal Hsp90 formulation, and Hsp90-expressing DNA vaccine, promote antibody-mediated protection against systemic candidiasis (Shukla et al. 2021).

**Monoclonal Antibodies**

Mycograb, often referred to as Efungumab, inhibits Hsp90 from producing antifungal effects. Clinical evidence backs the use of efungumab in conjunction with other antifungal medications to reduce invasive candidiasis (Karwa and Wargo 2009). In a neutropenic mouse model of systemic candidiasis, it was unexpectedly discovered that a variation of Mycograb provided no effective response when coupled with amphotericin B (Louie et al. 2011). The lack of both in vitro and in vivo efficacy rendered the combinatorial effect of the Mycograb C28Y variant in the enhancement of amphotericin-B response as nonspecific, in a mice model of candidiasis (Richie et al. 2012; Bugli et al. 2013).

## *Agglutinin-Like Sequence 3 (Als3)*

**Vaccine Research**

The Als3 encoded protein, which belongs to the family of agglutinin-like sequences, is agglutinin and is required for fungus invasion and adherence (Hoyer and Cota 2016). It triggers endocytosis and aids in fungus attachment to a variety of host tissues (Phan et al. 2007). Als3 was also discovered on hyphae obtained from a disseminated candidiasis model in mice (Coleman et al. 2009). Als3 is a key virulence factor since it has been demonstrated to be an invasin that may interact with *Candida* cell surface cadherins and cause fungal internalization through host cells (Coleman et al. 2009). The mutant and wild type parents' strains with Als3 mutation, however, remained equally virulent in a murine model of disseminated candidiasis (Cleary et al. 2011). Mutant Als3 *C. albicans* strains have been used to illustrate the function of Als3 in epithelial adherence, cell injury, cytokine generation, and activation of map-kinase-mediated signaling pathways (Murciano et al. 2012). According to a different study, the Als3 protein aids in the in vitro killing of fungi by oral and vaginal epithelial cells by encouraging the secretion of chemokines and cytokines during *Candida* infection (Gao et al. 2019). The mouse systemic candidiasis model can be utilized to assess the etiology and antifungal medication efficacy, according to a seminal work by Spellberg et al. (2005). An Als3 formulation plus alum as an adjuvant dramatically increased the survival of mice exposed to systemic *C. albicans* infection (Lin et al. 2008). According to Lin et al. research, IgG and IgG2a subclass antibodies made up most of the higher antibody titers that Als3 immunization in mice caused (Lin et al. 2009). According to Spellberg et al., protection against *Candida* infection was primarily provided by cell-mediated immunity, even though Als3 vaccination could generate primary B-cell response that resulted in higher levels of IgG as well as IgG2a antibody titers (Spellberg et al. 2008). This suggests that assessment of antibody response can be used as a substitute for indicators of vaccine-induced protection. Analysis of the immunological response of human blood samples to the recombinant Als3 vaccine revealed that serum levels of anti-Als3 IgG antibodies are detectable in healthy individuals (Baquir et al. 2010).

Schmidt et al. demonstrated that the Als3-based NDV-3 vaccine formulation can elicit a strong immune response in healthy humans (Schmidt et al. 2012). This immune response was shown by increased titers of anti-Als3 IgA1 and IgG antibodies. In contrast to the limited function identified for B cells in systemic candidiasis, NDV-3 immunization conferred effective protection in mice by generating both B and T cell-mediated immunity against vulvovaginal candidiasis (Ibrahim et al. 2013). Anti-Als3-N antibodies were passively transferred to naive mice, but this did not shield them from vaginal candidiasis (Ibrahim et al. 2013). Both subcutaneous and intramuscular vaccination resulted in greater anti-Als3 IgG antibody titers in vivo regardless of NDV3 dose (Ibrahim et al. 2013). Additionally, Yeaman et al. showed that by inducing a potent B and T-cell immune response, the Als3 vaccination may shield mice from infections with both methicillin-resistant *S. aureus* and *Candida* (Yeaman et al. 2014). Furthermore, a clinical evaluation of patients with previous episodes of vaginal candidiasis revealed that an Als3 with alum formulation (NDV-3A) was both immunogenic and safe. The quick and strong B and T-cell immune responses induced by the NDV-3A vaccine were observed to give protection against vaginal infections (Edwards et al. 2018). By increasing anti-Als3 antibody titers, which inhibit *C. albicans* from adhering, penetrating endothelial cells, and produce biofilms in vitro, it has been shown that NDV-3A immunization can confer protection mice from *C. albicans* infections (Alqarihi et al. 2019). Recently, it was discovered that NDV-3A immunization produces protective cross-reactive antibodies (Singh et al. 2019).

**Monoclonal Antibodies**

Anti-Als3 antibodies can interfere with *C. albicans* adhesion, filamentation, and biofilm formation, as reported by Uppuluri et al. (2018b). Als3 was found to bind to the monoclonal antibody C7, which has fungicidal activity and prevented fungi from adhering to epithelial surfaces and filamenting (Brena et al. 2007). Furthermore, adherence of *C. albicans* germ tubes to human epithelial as well as endothelial cells were found to be reduced in response to 3D9.3 monoclonal antibody, specific to Als3 protein (Beucher et al. 2009).

## *Secreted Aspartyl Proteinase 2 (Sap2)*

**Vaccine Research**

The secretory aspartyl proteinase (Sap) family, which consists of 10 members (Sap1-Sap10), includes the Sap2 gene (Naglik et al. 2003). Sap2 has been implicated in fungal virulence and is the predominant vaccine candidate derived from *C. albicans*. By breaking down various host proteins at epithelial cells and having the ability to hydrolyze complement, *Candida* Sap2 genes significantly contribute to fungal pathogenesis (Naglik et al. 2003).

The function of different Sap enzymes in the *Candida* pathogenicity has been confirmed through gene disruption studies, and Sap mutations resulted in a reduction in virulence during disseminated infections (Hube et al. 1997). Another work that used the triple Sap gene mutant (Sap4-6) to demonstrate the significance of Sap genes in *Candida* pathogenicity during murine systemic infection (Sanglard et al. 1997). In a rat model of vaginal candidiasis, De Bernardis et al. showed that *Candida* mutant strains missing Sap2 displayed decreased pathogenicity between Sap1 and Sap6 (De Bernardis et al. 1999). Prior research has shown that human antibody responses to *Candida* infection are produced by the *C. albicans* proteinases (Macdonald and Odds 1980; Borg and Ruchel 1988). In a rat model, Anti-Sap2 antibodies were originally shown to have a protective effect against *C. albicans*-mediated vaginal candidiasis (Cassone et al. 1995). According to De Bernardis et al., vaccination of rats against *Candida*-mediated vaginitis with anti-Sap2 monoclonal antibody, Sap2 antigen, and anti-Sap2 antibody from vaginal secretions provided protection (De Bernardis et al. 1997). Additionally, anti-Sap2 antibodies that were protective also showed reactivity with the other Sap proteins (White et al. 1993; White and Agabian 1995). Serum and saliva from HIV-infected individuals contained greater levels of total anti-Sap antibodies compared to controls (Drobacheff et al. 2001). In a rat model, anti-Sap2 antibodies were also reported to be protective against vulvovaginal candidiasis (Millon et al. 1999). A total of 6, Sap2-specific IgM and IgG B cell epitopes were discovered by Ghadjari et al. by analyzing sera from individuals suffering from oral and systemic candidiasis, and they may help to provide protection against disseminated candidiasis (Ghadjari et al. 1997).

In another study, the protective role of anti-Sap2 antibodies against *Candida* infection was reported (Vilanova et al. 2004). Mice protected against systemic candidiasis after receiving an injection of recombinant Sap2 protein, and kidney infection with *C. albicans* was considerably lessened by passive administration of anti-Sap2 antibody (IgG). The neutralization of enzymes is aided by the antibodies generated after Sap2 vaccination. In animal models, the aspartic protease inhibitor pepstatin therapy of mucosal and peritoneal *Candida* infections have been found to support the idea that Sap2 inhibition can reduce *Candida* infections (Staib et al. 2008). In mouse and rat models, De Bernardis et al. showed that intramuscular vaccination with a virosomal version of the Sap2 vaccine (PEV-7) causes an effective antibody response (De Bernardis et al. 2012). Rats immunized through intravaginal or intramuscular routes with recombinant Sap2 protein produced anti-Sap2 antibodies in vaginal fluids, while rats immunized with PEV7 through intravaginal or intramuscular methods displayed antibody-mediated protection against *C. albicans* vaginitis (De Bernardis et al. 2012). The PEV-7 preparation has successfully finished a Phase I clinical trial, and a significant humoral immune response was observed in vaginal as well as cervical samples after PEV-7 immunization using intramuscular injections or intravaginal capsules (De Bernardis et al. 2015). According to a study, BALB/c mice immunized with either recombinant Sap2 protein or a hybrid phage showing the Sap2 epitope SLAQVKYTSASSI elicited potent humoral and cellular responses against *C. albicans* infection (Wang et al. 2014). According to a recent study, Sap2 immunization led to higher levels of anti-Sap2 antibodies that bind the whole *Candida* yeasts (Shukla and Rohatgi 2020). Anti-Sap2 antibodies improved neutrophil-mediated fungal death, increased the ability of *Candida* to form biofilms in vitro, and shielded uninfected animals against systemic infection after passive administration. The results of this investigation also indicated a potential involvement for *Candida*-specific B1 and B2 B cells in the earlier phases of a Sap2-mediated immune response (Shukla and Rohatgi 2020).

**Monoclonal Antibodies**

The protective immune response against Sap2 was previously shown to be T-cell reliant and to be induced by anti-Sap2 antibodies. This was validated by pre-absorption of fluids containing Sap2 which decreased the degree of protection (Smolenski et al. 1996). Another study reported that Sap2 immunization conferred protection against *C. albicans* vaginitis by inducing elevated anti-Sap2 IgG and IgA antibody titers in rats (Sandini et al. 2011). Additionally, the protective efficacy of Sap2-specific antibodies were validated by further studies in which protection was mediated by passive administration of immunological vaginal fluids and anti-rSap2 IgG and IgM monoclonal antibodies (Sandini et al. 2011).

## *Hyphally Regulated Protein 1 (Hyr1)*

**Vaccine Research**

A hypha expressed gene called HYR1 (Hyphally regulated protein 1) is necessary for virulence and hyphal development. One group of researchers demonstrated that the Hyr1 protein was only expressed on the hyphae of *C. albicans* and had no impact on fungal germination (Bailey et al. 1996). Because Hyr1 is resistant to phagocyte death, a crucial component of the host's defense against candidiasis, Hyr1 is known to increase the pathogenicity of *C. albicans*. According to Luo et al., vaccination against disseminated candidiasis using rHyr1 protein along with alum adjuvant improved survival and decreased tissue fungal burden in both immunocompetent as well as immunocompromised mice (Luo et al. 2010).

Furthermore, Hyr1 protein was directly neutralized in vitro by passive administration with anti-Hyr1 protein polyclonal antibodies, increasing mouse neutrophil killing activity, protecting mice against *C. albicans* infection (Luo et al. 2010). A recombinant N-terminal region of *C. albicans* Hyr1 protein (rHyr1p-N) was used for mice vaccination. Mice were found to be protected against deadly candidemia by both active as well passive immunization with rHyr1p-N (Luo et al. 2011). The most plausible reason Hyr1 immunization protects against systemic candidiasis is because anti-Hyr1 antibodies directly promote neutrophil-mediated killing of pathogen (Cassone et al. 1930). According to Uppuluri et al., mice were protected from bacteremia and pneumonia caused by *Acinetobacter baumannii* by either an active immunization with rHyr1p or a passive administration with anti-Hyr1p antibodies (Uppuluri et al. 2018a). The formation of mixed biofilms by *C. albicans* and *A. baumannii* was also seen to be inhibited by polyclonal antibodies raised against Hyr1p-N peptides in vitro (Uppuluri et al. 2018a). Overall, above mentioned studies suggested that the Hyr1 vaccination strategy is reliant on the production of protective neutralizing antibodies.

**Monoclonal Antibodies**

In one study, a cross-kingdom immunity against gram-negative bacteria was offered by monoclonal IgM antibodies directed against the *C. albicans* Hyr1 protein (Youssef et al. 2020). From the serum of donors that was anti-*Candida* IgG-positive, single class switched memory B cells were isolated, which were further developed in vitro and tested against rHyr1 and entire fungal cell wall preparation (Rudkin et al. 2018). Monoclonal anti-*Candida* antibodies with the specificity for the human Hyr1 gene improved phagocytosis and provided protection from disseminated candidiasis (Rudkin et al. 2018).

## *Hyphal Wall Protein 1 (Hwp1)*

**Vaccine Research**

Hwp1 protein is encoded by Hwp1 gene, which is found on the hyphal cell wall of *C. albicans* and is a substrate for mammalian transglutaminase enzymes. It is an adhesion protein that aids in the fungal attachment with epithelial cells. It belongs to the family of GPI (glycosylphosphatidylinositol) anchor-dependent proteins, which is known to play a role in the creation of covalent connections with primary amines and buccal epithelial cells. It has been demonstrated that the Hwp1 protein is essential for *C. albicans* pathogenicity and in vivo hyphal growth (Tsuchimori et al. 2000).

Both in vitro and in vivo investigations have demonstrated that Hwp1 expression is essential for the development of biofilms, making it a viable therapeutic target (Nobile et al. 2006). According to research, the zinc finger protein transcription factor Bcr1 is necessary for Hwp1 expression which may aid in neutrophil evasion (Dwivedi et al. 2011). Naglik et al. observed that both candidiasis patients and healthy adults have systemic antibody-mediated responses to Hwp1 (IgG and IgA titers), indicating that Hwp1 plays a consistent role in the etiology of candidiasis (Naglik et al. 2006). Interesting evidence linking *C. albicans* infection and celiac disease has been found through humoral immunity, with celiac disease patients having high anti-Hwp1 antibody levels (Corouge et al. 2015).

**Monoclonal Antibodies**

Recently, Rosario-Colon et colleagues discovered that mice might be protected from invasive *C. auris* infection by *Candida* Hwp1-specific monoclonal antibodies by having higher survival rates and lower fungal loads (Rosario-colon et al. 2021).

## *Enolase (Eno)*

**Vaccine Research**

Enolase is a cytosolic enzyme involved in the glycolysis cycle that is widely expressed in metabolically active cells. Both on the surface of the fungal cell wall and in the extracellular media are where it is secreted (Silva et al. 2014). It facilitates fibrinolysis activation, extracellular matrix destruction, and fungal adherence to human tissues (Silva et al. 2014; Satala et al. 2020).

Anti-enolase antibodies have been demonstrated to prevent *C. albicans* from adhering to epithelial cells (Silva et al. 2014). Most studies from enolase-specific antibody responses imply that enolase-specific antibody production is a sensitive and early sign of *C. albicans* proliferation in mice. Antibody-mediated responses to enolase in humans are minimal or nonexistent during initial *Candida* colonization but become more pronounced following widespread infection in immunocompetent hosts. The immunogenicity of enolase was identified in a previous study which shows enolase to be an immunogenic enzyme (Sundstrom et al. 1994). In mice, systemic candidiasis was reported to be prevented by passive administration of anti-enolase antibodies (Van Deventer et al. 1996). Compared to either non-vaccinated or IL-12-treated mice, vaccination with a combination of enolase plus IL-12 displayed higher antibody titers against enolase, a longer median survival time, and lower kidney fungal load (Montagnoli et al. 2004). In spite of the significant immunogenicity shown for recombinant enolase, improved survival was seen in B-cell deficient mice, hence protection was suggested to be mainly associated to Th1 type response (Montagnoli et al. 2004). Recombinant enolase protein vaccination provided mice with an effective defense against widespread *C. albicans* infection, according to a different study. Enolase-specific antiserum that was passively transferred in mice which decreased the fungal growth in host tissue. Moreover, enolase-specific antibodies such as IgG1 and IgG2a antibodies could improve the neutrophil-mediated killing of *C. albicans* (Qing Li et al. 2011).

The serological identification of IgG antibodies against *Candida* enolase has been found associated with the diagnosis of systemic candidiasis (Laín et al. 2007; Li et al. 2013). One study reported that oral treatment of *Saccharomyces cerevisiae* cells bearing the enolase 1 antigen might trigger an immunological response and increase the survival rate of mice exposed to *C. albicans* infection (Shibasaki et al. 2013). Recently, a study examined the serological response to various *C. albicans* recombinant proteins such as β-glucosidase (rBgl2), phosphoglycerate kinase (rPgk1), and enolase (rEno1) in a mouse model of systemic candidiasis. Among the tested recombinant proteins, rEno1 demonstrated a stronger serological response than the rest of the two proteins (He et al. 2015).

**Monoclonal Antibodies**

Recently, Leu et al. employed phage display technique to identify a single chain variable fragment monoclonal antibody (CaS1) specific to the recombinant *C. albicans* enolase and also assessed its function in both in vitro and in vivo settings. CaS1 prevented *C. albicans* from growing and from binding to plasminogen in vitro. Further, in a mice model of systemic candidiasis, CaS1 injection increased survival time, decreased fungal loads, and decreased the levels of inflammatory cytokines (Leu et al. 2020). By neutralizing Eno1, the monoclonal antibody 12D9 (IgG) promoted opsonophagocytosis and neutrophil-mediated killing of *C. albicans* (Chen et al. 2020).

## *Phospholipase B (PLB)*

**Vaccine Research**

It has been suggested that phospholipases, a diverse group of enzymes that hydrolyze one or more ester bonds in glycerophospholipids, have a role in *C. albicans* pathogenicity. The enzyme phospholipase B (PLB), which can be identified in secreted as well as intracellular forms. It has both hydrolase and acyltransferase activity (Djordjevic 2010).

It is known to play a crucial part in cellular processes like inflammation and signal transduction by its impact on the phospholipid metabolism. Inactivating the PLB5 gene in *C. albicans* resulted in lowered activity of the phospholipase A2 enzyme which reduced the virulence of *Candida* (Theiss et al. 2006). PLB has been identified as a possible therapeutic target because of its confirmation as a virulence factor in disseminated candidiasis in animal models and its identification in other pathogenic fungi (Ghannoum 2000). Leidich et al. demonstrated PLB-deficient *Candida* strain's much poorer penetration of HUVEC and HT-29 epithelial cells in vitro compared to parent *Candida* strain (Leidich et al. 1998). These investigations suggest that *Candida* PLB may be essential for the spread of *C. albicans* via both gastrointestinal as well as hematogenous systems. Moreover, by reintroducing the PLB1 gene into *C. albicans,* the in vivo fungal pathogenicity could be restored (Mukherjee et al. 2001). Phospholipases may be employed as a candidate for a vaccine, according to the study, which used anti-PLB1 antibody to demonstrate that PLB1 is produced during invasion of the host gastric mucosal lining by *Candida* strains (Mukherjee et al. 2001).

Research has revealed that the serum of people with systemic candidiasis contains antibodies that react with purified *Candida* PLB, which makes it a promising option for diagnostic usage. Moreover, it has been shown to mediate B cell and antibody-mediated immune responses against *C. albicans*. Therefore, due to its *Candida* hyphae-specific property it can be explored as a vaccine candidate in future (Heilmann et al. 2011).

## *Fructose-Bisphosphate Aldolase (Fba1)*

**Vaccine Research**

A key enzyme in the glycolytic process and a multifunctional protein is fructose-bisphosphate aldolase (Fba1). It shields *Candida* cells from the immune system of the host and promotes fungal adhesion to human cells or abiotic surfaces (Elamin Elhasan et al. 2021). Furthermore, reactive oxygen species produced during respiratory burst are encouraged to be detoxified by Fba1. According to a proteome study, Fba1 is one of the highly abundant and persistent enzyme in *Candida* and is believed to be an important immunoreactive protein (Elamin Elhasan et al. 2021). In mice, immunization with the Fba peptide was found to stimulate the development of protective antibodies and protection against disseminated candidiasis (Xin et al. 2008). Later, the study also revealed that the combination generated protective antibody responses against peptide and glycan portions of the vaccine, which was supported by passive administration of immune sera in animal model (Xin et al. 2012). They used a 14-mer peptide of Fba protein to generate a self-adjuvanting vaccine. Fba peptide-pulsed-DC vaccination resulted in an elevated level of protective immunity against systemic candidiasis. Moreover, both active immunization with DC plus Fba peptide and passive administration of antibodies conferred protection in neutropenic mice (Xin 2016). In a mouse model, protection against systemic candidiasis was discovered to be conferred by phage vaccines showing the YGKDVKDLFDYAQE epitope from the Fba1 protein, mostly by activating adaptive immune responses.

The immunization dramatically increased the survival rates of sick mice by reducing the fungal burden and alleviating renal tissue damage (Shi et al. 2018). According to a study, recombinant Fba played an immunodominant role and vaccination with Fba-induced protection in mice against *C. glabrata*-mediated systemic infection (Medrano-Díaz et al. 2018). Additionally, vaccination with three mimotope–peptide combination vaccines proved effective in eliciting antibody responses and protecting mice from systemic candidiasis. Passive transfer experiments have shown antibody-mediated protection for Fba1 and found complement activation and impairment with hyphal development as defense mechanisms (Xin et al. 2019). Mice were vaccinated with a vaccine construct containing Fba and methionine synthase peptide (MP12), which protected them from systemic candidiasis by producing IgG1 and IgG2a antibodies, demonstrating the importance of humoral immune responses in the establishment of anti-*Candida* immunity (Adams et al. 2021). A recent study identified the most immunogenic and conserved B and T cell epitopes from the *C. glabrata* Fba1 protein which could be employed to develop an efficient epitope-based peptide vaccination (Elamin Elhasan et al. 2021).

**Monoclonal Antibodies**
Protective sera obtained from Fba-vaccinated mice was passively transferred to unvaccinated mice which conferred protection. An IgM monoclonal antibody specific to Fba peptide was generated and administered in mice. The monoclonal antibody conferred protection to mice which offered clear evidence that anti-Fba antibodies play an important role against systemic *Candida* infection (Xin and Cutler 2011).

## *Pyruvate Kinase (Pk)*

**Vaccine Research**
Pyruvate kinase (Pk), a glycolytic enzyme found in the cell walls of *C. albicans*, is of particular interest because it plays a role in the development of biofilms and oxidative stress. Pyruvate kinase and alcohol dehydrogenase were reported to be non-ubiquitous immunogenic proteins during *C. albicans* infections after searching for *C. albicans* sequences encoding for immunogenic proteins (Swoboda et al. 1993). Pk has been identified as one of the immunogenic proteins that induced the protective IgG2a antibody isotype to be produced in the serum of animals who had received vaccinations (Fernández-Arenas et al. 2004).

Pitarch and colleagues used two-dimensional polyacrylamide gel electrophoresis to analyze immunological serum from mice infected with *C. albicans* that were taken at various times after infection and found that pyruvate kinase was an immunoreactive antigen (Pitarch et al. 2001). In a more recent study, Medrano-Diaz et colleagues found that immunization with recombinant Pk antigen had the highest immunogenicity and could protect mice from systemic *C. albicans* infection (Medrano-Díaz et al. 2018).

## *Superoxide Dismutase (Sod5)*

**Vaccine Research**
Superoxide dismutase (Sod) is an antioxidant enzyme that changes superoxide radicals into hydrogen peroxide, which is less harmful. Six SOD gene family members

have been found in *C. albicans*. Sod5 is the only extracellular member and is connected to the fungal cell surface by GPI anchors (Schatzman et al. 2020). Sod5 is elevated during yeast-to-hyphae transitions, osmotic and oxidative stress conditions, and oxidative stress (Martchenko et al. 2004); it is also necessary for *C. albicans* pathogenicity in invasive animal models (König et al. 2020).

Sod5 is thought to be a viable candidate for a vaccination against systemic candidiasis since it mostly found as hyphae-associated cell wall protein (Gil-Bona et al. 2015). It was observed that the deletion of Sod5 gene caused a reduction in *C. albicans* virulence (Martchenko et al. 2004). Also, Sod5-deficient *C. albicans* mutants were highly vulnerable to neutrophil and macrophage-mediated killing, and they produced more reactive oxygen species in these cells (Frohner et al. 2009). Furthermore, one recent study showed that Sod5 gene deletion could prevent the development of *Candida* biofilm on venous catheters (Robinett et al. 2019). Moreover, it has been suggested that a novel therapeutic strategy to treat systemic *Candida* infection may involve inhibiting the extracellular portions of Sod enzymes of *C. albicans* (Frohner et al. 2009).

### *Malate Dehydrogenase (Mdh1)*

**Vaccine Research**

Few proteomics studies have looked at *C. albicans* malate dehydrogenase (Mdh1) as a potential target for a candidiasis vaccine (Fernández-Arenas et al. 2004; Aoki et al. 2013). The Mdh1 protein participates in the malate-aspartate shuttle, which is necessary for the TCA (tricarboxylic acid cycle) cycle to be completed and for aerobic energy production. Since Mdh1 was present at every time point examined without showing significant fluctuations in its relative abundance, it is recognized as a vaccination antigen against candidiasis (Shibasaki et al. 2018).

When recombinant Mdh1 protein was tested as an immunogenic target for generating a potent vaccine against candidiasis, subcutaneous and intradermal delivery of the protein greatly increased antibody responses and provided substantial protection against *C. albicans* in murine model of systemic candidiasis (Shibasaki et al. 2014). More such studies may be useful to develop an effective vaccine against systemic candidiasis.

## Conclusion

The assessment of host immune responses that guard against systemic *Candida* infections has made substantial progress over the past few decades. The fungal pathogen's cell wall contains a few virulent factors that aid in immune evasion. Fungal clearance requires the generation of efficient innate and adaptive immune responses.

The main components of innate immunity are phagocytes, complement proteins, and cytokines, while major players of adaptive immunity include T and B-cells. B-cells produce antibody-secreting plasma cells which are essential in anti-*Candida* protection. Even though cellular immunity has been extensively demonstrated in providing protection against systemic candidiasis. The role of humoral immune responses in triggering protective immunity against systemic candidiasis is still up for debate. Numerous *Candida* virulence antigens, such as cell surface proteins, polysaccharides, and enzymes, have been identified as potent vaccine candidates in diverse studies evaluating the defensive immune responses against *Candida* infection, mostly in murine models of systemic candidiasis. Utilizing Als3, Sap2, enolase, Hyr1, glucan and mannan, a variety of vaccines have been developed.

However, only the Als3 and Sap2 vaccines have completed phase-1 clinical trials, the remaining vaccines are under preclinical testing. Many antifungal vaccines confer protection to the host via activating humoral and/or cellular immune responses. Moreover, data from numerous literature reports that have been compiled in this chapter demonstrate unequivocally that antibodies produced in response to top vaccine candidates protect against systemic candidiasis through several different mechanisms. For instance, anti-Sap2 antibodies have characteristics including neutralization, preventing the growth of *Candida* biofilms, and enhancing neutrophil-mediated killing. Although T-cell-mediated immune responses are primarily responsible for Als3-mediated protection, primary B-cell responses result in elevated antibody titers, which are utilized as indicators of vaccine-induced protection. Anti-Hyr1 antibodies prevent biofilm development and play a direct, non-opsonic function in neutrophil-mediated fungal death, whereas anti-Als3 antibodies are known to inhibit *C. albicans* adhesion, filamentation, and biofilm formation.

Also, anti-enolase antibodies are known to prevent fungi from adhering to host cells and to improve opsonization-mediated *C. albicans* killing by neutrophils. By blocking the *Candida* Hsp90 antigen, which is essential for the survival of the fungus, antibodies against Hsp90 display direct anti-*Candida* activity. Anti-mannan antibodies can activate the complement system's alternative and classical pathways, enhancing the opsonophagocytic death of *C. albicans*. Antibodies against β-glucan and/or laminarin, which are known to prevent fungal growth and adhesion, have also been found to have direct antifungal activity. There are comparatively fewer data on how antibodies made against the vaccination antigens for PLB, Hwp1, Fba, Sod5, Pk, and Mdh1 work as a kind of protection. In addition to their role as APCs, B-cells can directly attach to *Candida* antigens and produce antibodies by differentiating into plasma cells, which are the major contributors in the antifungal humoral immunity.

Additionally, B-cells also form memory B-cells which shield against subsequent infections. Animals lacking B cells exhibit lower amounts of IL-4, TGF-β and IL-10 cytokines and are ineffective at developing a Th2-dependent immune response. Additionally, depleting B cells in vitro has been shown to have an antibody-independent role for B cells during a *Candida* infection. Additionally, it has been suggested that B1 B-cells play a role in innate immunity against systemic candidiasis.

The use of passive antibody treatment during disseminated candidiasis is supported by numerous studies in the literature, particularly in immunocompromised hosts. Animal research has provided compelling data, while clinical trials have not yet provided sufficient experimental support. However, human recombinant antibodies and anti-*Candida* monoclonal antibodies are regarded as promising immunotherapeutics for treating systemic candidiasis. Numerous anti-*Candida* immune responses, including neutralization, fungal growth suppression, opsonophagocytosis, direct candidacidal action and biofilm inhibition, have been demonstrated to be elicited by monoclonal antibodies produced against the most important vaccine antigens.

Additionally, it is known that anti-idiotypic antibodies and mimotopes that detect yeast-killing toxins have direct candidacidal activity. In numerous investigations, antibody-derived peptides and single chain variable fragments have been employed for either active or passive immunization in the animal models of experimental disseminated candidiasis. A compelling argument in favor of using antibodies as an adjuvant therapy is made by the discovery that anti-*Candida* antibodies directed against a single antigenic determinant such as β-glucan can show cross-reactivity and provide protective immunity against several fungal infections. In addition to offering a more robust and effective approach for addressing the recent prevalence of drug resistance in *Candida*, synergistic/combinatorial monoclonal antibodies and antifungal medications are also anticipated to benefit clinical results due to their increased specificity. Another possible strategy for increasing their therapeutic efficacy is the development of innovative immunomodulatory strategies that combine the administration of recombinant cytokines with the use of monoclonal antibodies. Additionally, cutting-edge technology like multivalent antibodies, antibody-antibiotic conjugates and antibody-drug conjugates, could be used to treat invasive fungal infections that are potentially fatal.

## Future Perspective

New therapeutic alternatives, such as vaccines and/or antibody-based therapeutics, are urgently needed to address the substantial threat posed by invasive fungal diseases. The necessity for developing alternative immunotherapies against systemic candidiasis is justified by the difficulties arising from the widespread use of antifungal medications, such as antifungal drug resistance, and antifungal drug associated toxicity.

Due to the increasing evidence pertaining to the potential of B cell and antibody-mediated immunity in imparting anti-*Candida* protection, the characterization of *Candida* antigens that induce protective antibodies will be vital for developing efficient multivalent or multi-epitope-based vaccines. Novel peptides and oligosaccharides must be created utilizing a combination of protective antigens to create multi-epitope conjugate vaccines that include B-cell and T-cell epitopes for memory responses. Even though there aren't many experimental vaccinations on the market, there isn't much research that uses animals. By identifying, isolating, and characterizing anti-*Candida* monoclonal antibodies that are protective against *Candida* infection, along with assessment of their mechanisms of protection is expected to aid in the development of more precise antibody-based therapeutics.

To support passive antibody treatment against systemic candidiasis, more clinical trials are required. For the treatment of invasive candidiasis, novel cytokine-based supplementary immunotherapies should be investigated. To show the viability of such immunotherapies for enhancing the prognosis of systemic candidiasis, more research is required. Monoclonal antibodies have been shown to provide synergistic protection against *Candida* when used with anti-fungal medications. To evaluate the effectiveness and safety of combination medicines, more research is required. Future studies should be conducted to characterize the protective B-lymphocyte and T-cell repertoires, assisting in the development of effective immunization techniques, given that immunocompromised persons with defects in immune cell repertoire are susceptible to systemic candidiasis.

## References

Adams AL, Eberle K, Colón JR, Courville E, Xin H (2021) Synthetic conjugate peptide Fba-Met6 (MP12) induces complement-mediated resistance against disseminated Candida albicans. Vaccine 39:4099–4107. https://doi.org/10.1016/j.vaccine.2021.06.001

Ahmad S, Alfouzan W (2021) Candida auris: epidemiology, diagnosis, pathogenesis, antifungal susceptibility, and infection control measures to combat the spread of infections in healthcare facilities. Microorganisms 9:807. https://doi.org/10.3390/microorganisms9040807

Alqarihi A, Singh S, Edwards JE, Ibrahim AS, Uppuluri P (2019) NDV-3A vaccination prevents C. albicans colonization of jugular vein catheters in mice. Sci Rep 9:1–6. https://doi.org/10.1038/s41598-019-42517-y

Aoki W, Ueda T, Tatsukami Y, Kitahara N, Morisaka H, Kuroda K, Ueda M (2013) Time-course proteomic profile of Candida albicans during adaptation to a fetal serum. Pathog Dis 67:67–75. https://doi.org/10.1111/2049-632X.12003

Ashman RB, Fulurija A, Papadimitriou JM (1999) Both CD4+ and CD8+ lymphocytes reduce the severity of tissue lesions in murine systemic candidiasis, and CD4+ cells also demonstrate strain-specific immunopathological effects. Microbiology 145:1631–1640. https://doi.org/10.1099/13500872-145-7-1631

Bailey DA, Feldmann PJF, Bovey M, Gow NAR, Brown AJP (1996) The Candida albicans HYR1 gene, which is activated in response to hyphal development, belongs to a gene family encoding yeast cell wall proteins. J Bacteriol 178:5353–5360. https://doi.org/10.1128/jb.178.18.5353-5360.1996

Balish E, Jensen J, Warner T (1993) Resistance of SCID mice to Candida albicans administered intravenously or colonizing the gut: role of polymorphonuclear leukocytes and macrophages. J Infect Dis 167:912–919. https://doi.org/10.1093/infdis/167.4.912

Baquir B, Lin L, Ibrahim AS, Fu Y, Avanesian V, Tu A, Edwards J Jr, Spellberg B (2010) Immunological reactivity of blood from healthy humans to the rAls3p-N vaccine protein. J Infect Dis 201:473–477. https://doi.org/10.1086/649901

Beucher B, Marot-Leblond A, Billaud-Nail S, Oh SH, Hoyer LL, Robert R (2009) Recognition of Candida albicans Als3 by the germ tube-specific monoclonal antibody 3D9.3. FEMS Immunol Med Microbiol 55:314–323

Bhattacharya S, Sae-Tia S, Fries BC (2020) Candidiasis and mechanisms of antifungal resistance. Antibiotics 9:1–19. https://doi.org/10.3390/antibiotics9060312
Bistoni F, Verducci G, Perito S, Vecchiarelli A, Puccetti P, Marconi P, Cassone A (1988) Immunomodulation by a low-virulence, agerminative variant of candida albicans. Further evidence for macrophage activation as one of the effector mechanisms of nonspecific anti-infectious protection. Med Mycol 26:285–299. https://doi.org/10.1080/02681218880000401
Bohler K, Klade H, Poitschek C, Reinthaller A (1994) Immunohistochemical study of in vivo and in vitro IgA coating of candida species in vulvovaginal candidiasis. Genitourin Med 70:182–186. https://doi.org/10.1136/sti.70.3.182
Borg M, Ruchel R (1988) Expression of extracellular acid proteinase by proteolytic Candida spp. during experimental infection of oral mucosa. Infect Immun 56:626–631
Boxx GM, Nishiya CT, Kozel TR, Zhang MX (2009) Characteristics of Fc-independent human antimannan antibody-mediated alternative pathway initiation of C3 deposition to Candida albicans. Mol Immunol 46:473–480. https://doi.org/10.1016/j.molimm.2008.10.008
Brena S, Omaetxebarría MJ, Elguezabal N, Cabezas J, Moragues MD, Pontón J (2007) Fungicidal monoclonal antibody C7 binds to Candida albicans Als3. Infect Immun 75:3680–3682. https://doi.org/10.1128/IAI.01840-06
Bromuro C, Torosantucci A, Chiani P, Conti S, Polonelli L, Cassone A (2002) Interplay between protective and inhibitory antibodies dictates the outcome of experimentally disseminated candidiasis in recipients of a Candida albicans vaccine. Infect Immun 70:5462–5470. https://doi.org/10.1128/IAI.70.10.5462-5470.2002
Bromuro C, Romano M, Chiani P, Berti F, Tontini M, Proietti D, Mori E, Torosantucci A, Costantino P, Rappuoli R et al (2010) Beta-glucan-CRM197 conjugates as candidates antifungal vaccines. Vaccine 28:2615–2623. https://doi.org/10.1016/j.vaccine.2010.01.012
Brown GD, Denning DW, Gow NAR, Levitz SM, Netea MG, White TC (2012) Hidden killers: human fungal infections. Sci Transl Med 4:165rv13. https://doi.org/10.1126/scitranslmed.3004404
Bugli F, Cacaci M, Martini C, Torelli R, Posteraro B, Sanguinetti M, Paroni Sterbini F (2013) Human monoclonal antibody-based therapy in the treatment of invasive candidiasis. Clin Dev Immunol 2013:403121. https://doi.org/10.1155/2013/403121
Caesar-TonThat TC, Cutler JE (1997) A monoclonal antibody to Candida albicans enhances mouse neutrophil candidacidal activity. Infect Immun 65:5354–5357. https://doi.org/10.1128/iai.65.12.5354-5357.1997
Calandra T, Roberts JA, Antonelli M, Bassetti M, Vincent JL (2016) Diagnosis and management of invasive candidiasis in the ICU: an updated approach to an old enemy. Crit Care 20:125. https://doi.org/10.1186/s13054-016-1313-6
Capodicasa C, Chiani P, Bromuro C, De Bernardis F, Catellani M, Palma AS, Liu Y, Feizi T, Cassone A, Benvenuto E et al (2011) Plant production of anti-β-glucan antibodies for immunotherapy of fungal infections in humans. Plant Biotechnol J 9:776–787. https://doi.org/10.1111/j.1467-7652.2010.00586.x
Carneiro C, Correia A, Lima T, Vilanova M, Pais C, Gomes AC, Real Oliveira MECD, Sampaio P (2016) Protective effect of antigen delivery using monoolein-based liposomes in experimental hematogenously disseminated candidiasis. Acta Biomater 39:133–145. https://doi.org/10.1016/j.actbio.2016.05.001
Carrow EW, Hector RF, Domer JE (1984) Immunodeficient CBA/N mice respond effectively to Candida albicans. Clin Immunol Immunopathol 33:371–380
Casadevall A, Feldmesser M, Pirofski LA (2002) Induced humoral immunity and vaccination against major human fungal pathogens. Curr Opin Microbiol 5:386–391. https://doi.org/10.1016/S1369-5274(02)00337-5
Cassone A, Bromuro C, Chiani P, Torosantucci A (1930) Hyr1 protein and β-glucan conjugates as anti-candida vaccines. J Infect Dis 2010:202. https://doi.org/10.1086/657417
Cassone A, Boccanera M, Adriani D, Santoni G, De Bernardis F (1995) Rats clearing a vaginal infection by Candida albicans acquire specific, antibody-mediated resistance to vaginal reinfection. Infect Immun 63:2619–2624. https://doi.org/10.1128/iai.63.7.2619-2624.1995

Cassone A, Bernardis F, Torososantucci A (2005) An outline of the role of anti-Candida antibodies within the context of passive immunization and protection from candidiasis. Curr Mol Med 5:377–382. https://doi.org/10.2174/1566524054022549

Cassone A, De Bernardis F, Santoni G (2007) Anticandidal immunity and vaginitis: novel opportunities for immune intervention. Infect Immun 75:4675–4686. https://doi.org/10.1128/IAI.00083-07

Chen SM, Zou Z, Guo SY, Hou WT, Qiu XR, Zhang Y, Song LJ, Hu XY, Jiang YY, Shen H et al (2020) Preventing Candida albicans from subverting host plasminogen for invasive infection treatment. Emerg Microbes Infect 9:2417–2432. https://doi.org/10.1080/22221751.2020.1840927

Chiani P, Bromuro C, Cassone A, Torosantucci A (2009) Anti-β-glucan antibodies in healthy human subjects. Vaccine 27:513–519. https://doi.org/10.1016/j.vaccine.2008.11.030

Cleary IA, Reinhard SM, Lindsay Miller C, Murdoch C, Thornhill MH, Lazzell AL, Monteagudo C, Thomas DP, Saville SP (2011) Candida albicans adhesin Als3p is dispensable for virulence in the mouse model of disseminated candidiasis. Microbiology 157:1806–1815. https://doi.org/10.1099/mic.0.046326-0

Coleman DA, Oh SH, Zhao X, Zhao H, Hutchins JT, Vernachio JH, Patti JM, Hoyer LL (2009) Monoclonal antibodies specific for Candida albicans Als3 that immunolabel fungal cells in vitro and in vivo and block adhesion to host surfaces. J Biotechnol Methods 78:71–78. https://doi.org/10.1016/j.mimet.2009.05.002.Monoclonal

Conti HR, Gaffen SL (2010) Host responses to Candida albicans: Th17 cells and mucosal candidiasis. Microbes Infect 12:518–527. https://doi.org/10.1016/j.micinf.2010.03.013

Corouge M, Loridant S, Fradin C, Salleron J, Damiens S, Moragues MD, Souplet V, Jouault T, Robert R, Dubucquoi S et al (2015) Humoral immunity links Candida albicans infection and celiac disease. PLoS One 10:1–17. https://doi.org/10.1371/journal.pone.0121776

Cutler J (2005) Defining criteria for anti-mannan antibodies to protect against candidiasis. Curr Mol Med 5:383–392. https://doi.org/10.2174/1566524054022576

Daniel Gozalbo MLG (2009) IFN-g in Candida albicans infections. Front Biosci 14:1970–1978

De Bernardis F, Boccanera M, Adriani D, Spreghini E, Santoni G, Cassone A (1997) Protective role of antimannan and anti-aspartyl proteinase antibodies in an experimental model of Candida albicans vaginitis in rats. Infect Immun 65:3399–3405. https://doi.org/10.1128/iai.65.8.3399-3405.1997

De Bernardis F, Arancia S, Morelli L, Hube B, Sanglard D, Schäfer W, Cassone A (1999) Evidence that members of the secretory aspartyl proteinase gene family, in particular SAP2, are virulence factors for Candida vaginitis. J Infect Dis 179:201–208

De Bernardis F, Liu H, O'Mahony R, La Valle R, Bartollino S, Sandini S, Grant S, Brewis N, Tomlinson I, Basset RC et al (2007) Human domain antibodies against virulence traits of Candida albicans inhibit fungus adherence to vaginal epithelium and protect against experimental vaginal candidiasis. J Infect Dis 195:149–157. https://doi.org/10.1086/509891

De Bernardis F, Santoni G, Boccanera M, Lucciarini R, Arancia S, Sandini S, Amantini C, Cassone A (2010) Protection against rat vaginal candidiasis by adoptive transfer of vaginal B lymphocytes. FEMS Yeast Res 10:432–440. https://doi.org/10.1111/j.1567-1364.2010.00620.x

De Bernardis F, Amacker M, Arancia S, Sandini S, Gremion C, Zurbriggen R, Moser C, Cassone A (2012) A virosomal vaccine against candidal vaginitis: immunogenicity, efficacy and safety profile in animal models. Vaccine 30:4490–4498. https://doi.org/10.1016/j.vaccine.2012.04.069

De Bernardis F, Arancia S, Sandini S, Graziani S, Norelli S (2015) Studies of immune responses in Candida vaginitis. Pathogens 4:697–707. https://doi.org/10.3390/pathogens4040697

de Carvalho RJ, Cunha CM, Silva DA, Sopelete MC, Urzedo JE, Moreira TA, Moraes Pde S, Taketomi EA (2003) IgA, IgE and IgG subclasses to Candida albicans in serum and vaginal fluid from patients with vulvovaginal candidiasis. Rev Assoc Med Bras 49:434–438

Djordjevic JT (2010) Role of phospholipases in fungal fitness, pathogenicity, and drug development – lessons from Cryptococcus neoformans. Front Microbiol 1:25. https://doi.org/10.3389/fmicb.2010.00125

Drobacheff C, Millon L, Monod M, Piarroux R, Robinet E, Laurent R, Meillet D (2001) Increased serum and salivary immunoglobulins against Candida albicans in HIV-infected patients with oral candidiasis. Clin Chem Lab Med 39:519–526

Dwivedi P, Thompson A, Xie Z, Kashleva H, Ganguly S, Mitchell AP, Dongari-Bagtzoglou A (2011) Role of Bcr1-activated genes Hwp1 and Hyr1 in Candida albicans oral mucosal biofilms and neutrophil evasion. PLoS One 6:e16218. https://doi.org/10.1371/journal.pone.0016218

Edwards JE, Schwartz MM, Schmidt CS, Sobel JD, Nyirjesy P, Schodel F, Marchus E, Lizakowski M, Demontigny EA, Hoeg J et al (2018) A fungal immunotherapeutic vaccine (NDV-3A) for treatment of recurrent vulvovaginal candidiasis-a phase 2 randomized, double-blind, placebo-controlled trial. Clin Infect Dis 66:1928–1936. https://doi.org/10.1093/cid/ciy185

Elamin Elhasan LM, Hassan MB, Elhassan RM, Abdelrhman FA, Salih EA, Ibrahim HA, Mohamed AA, Osman HS, Khalil MS, Alsafi AA et al (2021) Epitope-based peptide vaccine design against fructose bisphosphate aldolase of Candida glabrata: an immunoinformatics approach. J Immunol Res 2021:1–19. https://doi.org/10.1155/2021/8280925

Fernández-Arenas E, Molero G, Nombela C, Diez-Orejas R, Gil C (2004) Low virulent strains of Candida albicans: unravelling the antigens for a future vaccine. Proteomics 4:3007–3020

François LM, Duncan W, Bernhard H (2013) Candida albicans pathogenicity mechanisms. Virulence 4:119–128. https://doi.org/10.4161/viru.22913

Frohner IE, Bourgeois C, Yatsyk K, Majer O, Kuchler K (2009) Candida albicans cell surface superoxide dismutases degrade host-derived reactive oxygen species to escape innate immune surveillance. Mol Microbiol 71:240–252. https://doi.org/10.1111/j.1365-2958.2008.06528.x

Gao Y, Liang G, Wang Q, She X, Shi D, Shen Y, Su X, Wang X, Wang W, Li D et al (2019) Different host immunological response to C. albicans by human oral and vaginal epithelial cells. Mycopathologia 184:1–12. https://doi.org/10.1007/s11046-018-0301-6

Gardner A (2007) Diagnosing fungal infections in neutropenic patients. Clin J Oncol Nurs 11:29–32

Ghadjari A, Matthews RC, Burnie JP (1997) Epitope mapping Candida Albicans proteinase (SAP 2). FEMS Immunol Med Microbiol 19:115–123

Ghannoum MA (2000) Potential role of phospholipases in virulence and fungal pathogenesis. Clin Microbiol Rev 13:122–143. https://doi.org/10.1128/CMR.13.1.122-143.2000

Gil-Bona A, Llama-Palacios A, Parra CM, Vivanco F, Nombela C, Monteoliva L, Gil C (2015) Proteomics unravels extracellular vesicles as carriers of classical cytoplasmic proteins in Candida albicans. J Proteome Res 14:142–153. https://doi.org/10.1021/pr5007944

Gintjee TJ, Donnelley MA, Thompson GR (2020) Aspiring antifungals: review of current antifungal pipeline developments. J Fungi 6:28. https://doi.org/10.3390/jof6010028

Hamad M (2012) Innate and adaptive antifungal immune responses: partners on an equal footing. Mycoses 55:205–217. https://doi.org/10.1111/j.1439-0507.2011.02078.x

Han Y, Cutler JE (1995) Antibody response that protects against disseminated candidiasis. Infect Immun 63:2714–2719. https://doi.org/10.1128/iai.63.7.2714-2719.1995

Han Y, Kanbe T, Cherniak R, Cutler JE (1997) Biochemical characterization of Candida albicans epitopes that can elicit protective and nonprotective antibodies. Infect Immun 65:4100–4107. https://doi.org/10.1128/iai.65.10.4100-4107.1997

Han Y, Morrison RP, Cutler JE (1998) A vaccine and monoclonal antibodies that enhance mouse resistance to Candida albicans vaginal infection. Infect Immun 66:5771–5776. https://doi.org/10.1128/iai.66.12.5771-5776.1998

Han Y, Riesselman MH, Cutler JE (2000) Protection against candidiasis by an immunoglobulin G3 (IgG3) monoclonal antibody specific for the same mannotriose as an IgM protective antibody. Infect Immun 68:1649–1654. https://doi.org/10.1128/IAI.68.3.1649-1654.2000

Han Y, Kozel TR, Zhang MX, MacGill RS, Carroll MC, Cutler JE (2001) Complement is essential for protection by an IgM and an IgG3 monoclonal antibody against experimental, hematogenously disseminated candidiasis. J Immunol 167:1550–1557. https://doi.org/10.4049/jimmunol.167.3.1550

Harpf V, Rambach G, Würzner R, Lass-Flörl C, Speth C (2020) Candida and complement: new aspects in an old battle. Front Immunol 11:1–13. https://doi.org/10.3389/fimmu.2020.01471
He ZX, Chen J, Li W, Cheng Y, Zhang HP, Zhang LN, Hou TW (2015) Serological response and diagnostic value of recombinant candida cell wall protein enolase, phosphoglycerate kinase, and β-glucosidase. Front Microbiol 6:1–9. https://doi.org/10.3389/fmicb.2015.00920
Heilmann CJ, Sorgo AG, Siliakus AR, Dekker HL, Brul S, Koster CG, de Koning LJ, Klis FM (2011) Hyphal induction in the human fungal pathogen Candida albicans reveals a characteristic wall protein profile. Microbiology 157:2297–2307. https://doi.org/10.1099/mic.0.049395-0
Hoogeboom R, van Kessel KP, Hochstenbach F, Wormhoudt TA, Reinten RJ, Wagner K, van Noesel CJ (2013) A mutated B cell chronic lymphocytic leukemia subset that recognizes and responds to fungi. J Exp Med 210(1):59–70
Hoyer LL, Cota E (2016) Candida albicans agglutinin-like sequence (Als) family vignettes: a review of Als protein structure and function. Front Microbiol 7:280
Hube B, Sanglard D, Odds FC, Hess D, Brown AJP, Gow NAR, Monod M, Scha W (1997) Disruption of each of the secreted aspartyl proteinase genes attenuates virulence. Infect Immun 65:3529–3538
Ibrahim AS, Luo G, Gebremariam T, Lee H, Schmidt CS, Hennessey JP, French SW, Yeaman MR, Filler SG, Edwards JE (2013) NDV-3 protects mice from vulvovaginal candidiasis through T- and B-cell immune response. Vaccine 31:5549–5556. https://doi.org/10.1016/j.vaccine.2013.09.016
Ishibashi KI, Yoshida M, Nakabayashi I, Shinohara H, Miura NN, Yoshiyuki Adachi NO (2005) Role of anti-beta-glucan antibody in host defense against fungi. FEMS Immunol Med Microbiol 44:99–109
Ishibashi KI, Dogasaki C, Motoi M, Miura N, Adachi Y, Ohno N (2010) Anti-fungal cell wall β-glucan antibody in animal sera. Japanese J Med Mycol 51:99–107. https://doi.org/10.3314/jjmm.51.99
Jones-Carson J, Vazquez-Torres FA, Balish E (1997) B cell-independent selection of memory T cells after mucosal immunization with Candida albicans. J Immunol 158:4328–4335
Kabir ME, Karim N, Krishnaswamy S, Selvakumar D, Miyamoto M, Furuichi Y, Komiyama T (2011) Peptide derived from anti-idiotypic single-chain antibody is a potent antifungal agent compared to its parent fungicide HM-1 killer toxin peptide. Applied microbiology and biotechnology 92(6):1151–1160
Karwa R, Wargo KA (2009) Efungumab: a novel agent in the treatment of invasive candidiasis. Ann Pharmacother 43:1818–1823. https://doi.org/10.1345/aph.1M218
König A, Hube B, Kasper L (2020) The dual function of the fungal toxin Candidalysin during Candida albicans-macrophage interaction and virulence. Toxins (Basel) 12:469
Kozel TR, MacGill RS, Percival A, Zhou Q (2004) Biological activities of naturally occurring antibodies reactive with Candida albicans Mannan. Infect Immun 72:209–218. https://doi.org/10.1128/IAI.72.1.209-218.2004
Kullberg BJ, Arendrup MC (2015) Invasive Candidiasis. N Engl J Med 373:1445–1456. https://doi.org/10.1056/NEJMra1315399
Laín A, Moragues MD, García Ruiz JC, Mendoza J, Camacho A, Del Palacio A, Pontón J (2007) Evaluation of a novel enzyme-linked immunosorbent assay to detect immunoglobulin G antibody to enolase for serodiagnosis of invasive candidiasis. Clin Vaccine Immunol 14:318–319. https://doi.org/10.1128/CVI.00396-06
Leidich SD, Ibrahim AS, Fu Y, Koul A, Jessup C, Vitullo J, Fonzi W, Mirbod F, Nakashima S, Nozawa Y et al (1998) Cloning and disruption of caPLB1, a phospholipase B gene involved in the pathogenicity of Candida albicans. J Biol Chem 273:26078–26086. https://doi.org/10.1074/jbc.273.40.26078
Leu SJ, Lee YC, Lee CH, Liao PY, Chiang CW, Yang CM, Su CH, Ou TY, Liu KJ, Lo HJ et al (2020) Generation and characterization of single chain variable fragment against alpha-enolase of Candida albicans. Int J Mol Sci 21:2903. https://doi.org/10.3390/ijms21082903

Li W, Fu M, An JG, Xing Y, Zhang P, Zhang X, Wang YC, Li CX, Tian R, Su WJ et al (2007) Host defence against C. albicans infections in IgH transgenic mice with VH derived from a natural anti-keratin antibody. Cell Microbiol 9:306–315. https://doi.org/10.1111/j.1462-5822.2006.00786.x

Li FQ, Ma CF, Shi LN, Lu JF, Wang Y, Huang M, Kong QQ (2013) Diagnostic value of immunoglobulin G antibodies against Candida enolase and fructose-bisphosphate aldolase for candidemia. BMC Infect Dis 13:1. https://doi.org/10.1186/1471-2334-13-253

Li R, Rezk A, Li H, Gommerman JL, Prat A, Bar-Or A (2017) Antibody-independent function of human B cells contributes to antifungal T cell responses. J Immunol 198:3245–3254. https://doi.org/10.4049/jimmunol.1601572

Lilly EA, Ikeh M, Nash EE, Fidel PL Jr, Noverr MC (2018) Immune protection against lethal fungal-bacterial intra-abdominal infections. MBio 9:1–14

Lin L, Ibrahim AS, Avanesian V, Edwards JE, Fu Y, Baquir B, Taub R, Spellberg B (2008) Considerable differences in vaccine immunogenicities and efficacies related to the diluent used for aluminum hydroxide adjuvant. Clin Vaccine Immunol 15:582–584. https://doi.org/10.1128/CVI.00427-07

Lin L, Ibrahim AS, Baquir B, Avanesian V, Fu Y, Spellberg B (2009) Immunological surrogate marker of rAls3p-N vaccine-induced protection against Staphylococcus aureus. FEMS Immunol Med Microbiol 55:293–295

Lionakis MS (2014) New insights into innate immune control of systemic candidiasis. Med Mycol 52:555–564. https://doi.org/10.1093/mmy/myu029

Lionakis MS, Netea MG (2013) Candida and host determinants of susceptibility to invasive candidiasis. PLoS Pathog 9:e1003079. https://doi.org/10.1371/journal.ppat.1003079

Lipinski T, Kitov PI, Szpacenko A, Paszkiewicz E, Bundle DR (2011) Synthesis and immunogenicity of a glycopolymer conjugate. Bioconjug Chem 22:274–281. https://doi.org/10.1021/bc100397b

Lipinski T, Wu X, Sadowska J, Kreiter E, Yasui Y, Cheriaparambil S, Rennie R, Bundle DR (2012) A β-mannan trisaccharide conjugate vaccine aids clearance of Candida albicans in immunocompromised rabbits. Vaccine 30:6263–6269. https://doi.org/10.1016/j.vaccine.2012.08.010

Lopez-Ribot Jose L, Casanova M, Murgui A, Martinez J (2004) Antibody response to Candida albicans cell wall antigens. FEMS Immunol Med Microbiol 41:187–196. https://doi.org/10.1016/j.femsim.2004.03.012

Louie A, Stein DS, Zack JZ, Liu W, Conde H, Fregeau C, VanScoy BD, Drusano GL (2011) Dose range evaluation of mycograb C28Y variant, a human recombinant antibody fragment to heat shock protein 90, in combination with amphotericin B-desoxycholate for treatment of murine systemic candidiasis. Antimicrob Agents Chemother 55:3295–3304. https://doi.org/10.1128/AAC.01324-10

Luo G, Ibrahim AS, Spellberg B, Nobile CJ, Mitchell AP, Fu Y (2010) Candida albicans Hyr1p confers resistance to neutrophil killing and is a potential vaccine target. J Infect Dis 201:1718–1728. https://doi.org/10.1086/652407.Candida

Luo G, Ibrahim AS, French SW, Edwards JE, Fu Y (2011) Active and passive immunization with rHyr1p-N protects mice against hematogenously disseminated candidiasis. PLoS One 6:e25909. https://doi.org/10.1371/journal.pone.0025909

Macdonald F, Odds FC (1980) Inducible proteinase of Candida albicans in diagnostic serology and in the pathogenesis of systemic candidosis. J Med Microbiol 13:423–435

Machová E, Korcová J, Čížová A, Bystrický S (2015) Inhibition of yeast growth by broadly cross-reactive antisera elicited by heterologous mannan-protein conjugate. J Microbiol Biotechnol 25:1177–1179. https://doi.org/10.4014/jmb.1410.10050

Magliani W, Conti S, Giovati L, Maffei DL, Polonelli L (2008) Anti-beta-glucan-like immunoprotective candidacidal antiidiotypic antibodies. Front Biosci 13:6920–6937

Maiti PK, Kumar A, Kumar R, Mohapatra LN (1985) Role of antibodies and effect of BCG vaccination in experimental candidiasis in mice. Mycopathologia 91:79–85. https://doi.org/10.1007/BF00436540

Mangeney M, Fischer A, Le Deist F, Latgé JP, Durandy A (1989) Direct activation of human B lymphocytes by Candida albicans-derived mannan antigen. Cell Immunol 122:329–337. https://doi.org/10.1016/0008-8749(89)90081-6

Martchenko M, Alarco AM, Harcus D, Whiteway M (2004) Superoxide dismutases in Candida albicans: transcriptional regulation and functional characterization of the hyphal-induced SOD5 gene. Mol Biol Cell 15:456–467

Matthews R, Burnie J (1992) The role of hsp90 in fungal infection. Immunol Today 13:345–348

Matthews R, Burnie JP, Lee W (1991a) The application of epitope mapping in the development of a new serological test for systemic candidosis. J Immunol Methods 143:73–79. https://doi.org/10.1016/0022-1759(96)00071-3

Matthews RC, Burnie JP, Howat D, Rowland T, Walton F (1991b) Autoantibody to heat-shock protein 90 can mediate protection against systemic candidosis. Immunology 74:20–24

Matthews R, Hodgetts S, Burnie J (1995) Preliminary assessment of a human recombinant antibody fragment to hsp90 in murine invasive candidiasis. J Infect Dis 171:1668–1671

Matveev AL, Krylov VB, Khlusevich YA, Baykov IK, Yashunsky DV, Emelyanova LA, Nifantiev NE (2019) Novel mouse monoclonal antibodies specifically recognizing β-(1→ 3)-D-glucan antigen. PLoS One 14(4):e0215535

Medici NP, Del Poeta M (2015) New insights on the development of fungal vaccines: from immunity to recent challenges. Mem Inst Oswaldo Cruz 110:966–973. https://doi.org/10.1590/0074-02760150335

Medrano-Díaz CL, Vega-González A, Ruiz-Baca E, Moreno A, Cuéllar-Cruz M (2018) Moonlighting proteins induce protection in a mouse model against Candida species. Microb Pathog 124:21–29. https://doi.org/10.1016/j.micpath.2018.08.024

Mikulska M, Calandra T, Sanguinetti M, Poulain D, Viscoli C (2010) The use of mannan antigen and anti-mannan antibodies in the diagnosis of invasive candidiasis: recommendations from the Third European Conference on Infections in Leukemia. Crit Care 14:1–14. https://doi.org/10.1186/cc9365

Millon L, Drobacheff C, Piarroux R, Monod M, Reboux G, Laurent R, Meillet D (1999) Longitudinal study of anti-Candida albicans mucosal immunity against aspartic proteinases in HIV-infected patients. J Acquir Immune Defic Syndr 2001(26):137–144

Montagnoli C, Sandini S, Bacci A, Romani L, La Valle R (2004) Immunogenicity and protective effect of recombinant enolase of Candida albicans in a murine model of systemic candidiasis. Med Mycol 42:319–324

Mukherjee PK, Seshan KR, Leidich SD, Chandra J, Cole GT, Ghannoum MA (2001) Reintroduction of the PLB1 gene into Candida albicans restores virulence in vivo. Microbiology 147:2585–2597. https://doi.org/10.1099/00221287-147-9-2585

Murciano C, Moyes DL, Runglall M, Tobouti P, Islam A, Hoyer LL, Naglik JR (2012) Evaluation of the role of Candida albicans agglutinin-like sequence (ALS) proteins in human oral epithelial cell interactions. PLoS One 7:1–9. https://doi.org/10.1371/journal.pone.0033362

Naglik JR, Challacombe SJ, Hube B (2003) Candida albicans secreted aspartyl proteinases in virulence and pathogenesis. Microbiol Mol Biol Rev 67:400–428. https://doi.org/10.1128/MMBR.67.3.400

Naglik JR, Fostira F, Ruprai J, Staab JF, Challacombe SJ, Sundstrom P (2006) Candida albicans HWP1 gene expression and host antibody responses in colonization and disease. J Med Microbiol 55:1323–1327. https://doi.org/10.1099/jmm.0.46737-0

Nami S, Aghebati-Maleki A, Morovati H, Aghebati-Maleki L (2019) Current antifungal drugs and immunotherapeutic approaches as promising strategies to treatment of fungal diseases. Biomed Pharmacother 110:857–868. https://doi.org/10.1016/j.biopha.2018.12.009

Netea MG, van Tits LJ, Curfs JH, Amiot F, Meis JF, van der Meer JW, Kullberg BJ (1999) Increased susceptibility of TNF-alpha lymphotoxin-alpha double knockout mice to systemic candidiasis through impaired recruitment of neutrophils and phagocytosis of Candida albicans. J Immunol 163:1498–1505

Netea MG, Gijzen K, Coolen N, Verschueren I, Figdor C, Van Der Meer JWM, Torensma R, Kullberg BJ (2004) Human dendritic cells are less potent at killing Candida albicans than both monocytes and macrophages. Microbes Infect 6:985–989. https://doi.org/10.1016/j.micinf.2004.05.013

Netea MG, Joosten LAB, Van Der Meer JWM, Kullberg BJ, Van De Veerdonk FL (2015) Immune defence against Candida fungal infections. Nat Rev Immunol 15:630–642. https://doi.org/10.1038/nri3897

Newman SL, Holly A (2001) Candida albicans is phagocytosed, killed, and processed for antigen presentation by human dendritic cells. Infect Immun 69:6813–6822. https://doi.org/10.1128/IAI.69.11.6813-6822.2001

Ngo LY, Kasahara S, Kumasaka DK, Knoblaugh SE, Jhingran A, Hohl TM (2014) Inflammatory monocytes mediate early and organ-specific innate defense during systemic candidiasis. J Infect Dis 209:109–119. https://doi.org/10.1093/infdis/jit413

Nishiya CT, Boxx GM, Robison K, Itatani C, Kozel TR, Zhang MX (2016) Influence of IgG subclass on human antimannan antibody-mediated resistance to hematogenously disseminated candidiasis in mice. Infect Immun 84:386–394. https://doi.org/10.1128/IAI.00890-15

Nobile CJ, Andes DR, Nett JE, Smith FJ, Yue F, Phan QT, Edwards JE, Filler SG, Mitchell AP (2006) Critical role of Bcr1-dependent adhesins in C. albicans biofilm formation in vitro and in vivo. PLoS Pathog 2:0636–0649. https://doi.org/10.1371/journal.ppat.0020063

Papon N, Courdavault V, Clastre M, Bennett RJ, Nac N-C (2013) Emerging and emerged pathogenic candida species: beyond the Candida albicans paradigm. J Patient Saf Infect Control 9:e1003550. https://doi.org/10.1371/journal.ppat.1003550

Pappas PG, Lionakis MS, Arendrup MC, Ostrosky-Zeichner L, Kullberg BJ (2018) Invasive candidiasis. Nat Rev Dis Prim 4:1–20. https://doi.org/10.1038/nrdp.2018.26

Paulovičová E, Machová E, Tulinská J, Bystrický S (2007) Cell and antibody mediated immunity induced by vaccination with novel Candida dubliniensis mannan immunogenic conjugate. Int Immunopharmacol 7:1325–1333. https://doi.org/10.1016/j.intimp.2007.05.014

Paulovičová L, Bystrický S, Paulovičová E, Karelin AA, Tsvetkov YE, Nifantiev NE (2010) Model α-mannoside conjugates: immunogenicity and induction of candidacidal activity. FEMS Immunol Med Microbiol 58:307–313. https://doi.org/10.1111/j.1574-695X.2009.00642.x

Phan QT, Myers CL, Fu Y, Sheppard DC, Yeaman MR, Welch WH, Ibrahim AS, Edwards JE, Filler SG (2007) Als3 is a Candida albicans invasin that binds to cadherins and induces endocytosis by host cells. PLoS Biol 5:0543–0557. https://doi.org/10.1371/journal.pbio.0050064

Pietrella D, Rachini A, Torosantucci A, Chiani P, Brown AJP, Bistoni F, Costantino P, Mosci P, d'Enfert C, Rappuoli R et al (2010) A β-glucan-conjugate vaccine and anti-β-glucan antibodies are effective against murine vaginal candidiasis as assessed by a novel in vivo imaging technique. Vaccine 28:1717–1725. https://doi.org/10.1016/j.vaccine.2009.12.021

Pitarch A, Díez-Orejas R, Molero G, Pardo M, Sánchez M, Gil C, Nombela C (2001) Analysis of the serologic response to systemic Candida albicans infection in a murine model. Proteomics 1:550–559. https://doi.org/10.1002/1615-9861(200104)1:4<550::AID-PROT550>3.0.CO;2-W

Polonelli L, Magliani W, Conti S, Bracci L, Lozzi L, Neri P, Adriani D, De Bernardis F, Cassone A (2003) Therapeutic activity of an engineered synthetic killer antiidiotypic antibody fragment against experimental mucosal and systemic candidiasis. Infect Immun 71:6205–6212. https://doi.org/10.1128/IAI.71.11.6205-6212.2003

Qing Li W, Chu Hu X, Zhang X, Ge Y, Zhao S, Hu Y, Ashman RB (2011) Immunisation with the glycolytic enzyme enolase confers effective protection against Candida albicans infection in mice. Vaccine 29:5526–5533. https://doi.org/10.1016/j.vaccine.2011.05.030

Quintin J, Voigt J, van der Voort R, Jacobsen ID, Verschueren I, Hube B, Giamarellos-Bourboulis EJ, van der Meer JWM, Joosten LAB, Kurzai O et al (2014) Differential role of NK cells against Candida albicans infection in immunocompetent or immunocompromised mice. Eur J Immunol 44:2405–2414. https://doi.org/10.1002/eji.201343828

Rapaka RR, Ricks DM, Alcorn JF, Chen K, Khader SA, Zheng M, Plevy S, Bengtén E, Kolls JK (2010) Conserved natural IgM antibodies mediate innate and adaptive immunity against the opportunistic fungus pneumocystis murina. J Exp Med 207:2907–2919. https://doi.org/10.1084/jem.20100034

Richie DL, Ghannoum MA, Isham N, Thompson KV, Ryder NS (2012) Nonspecific effect of mycograb on amphotericin B MIC. Antimicrob Agents Chemother 56:3963–3964. https://doi.org/10.1128/AAC.00435-12

Robinett NG, Culbertson EM, Peterson RL, Sanchez H, Andes DR, Nett JE, Culotta VC (2019) Exploiting the vulnerable active site of a copper-only superoxide dismutase to disrupt fungal pathogenesis. J Biol Chem 294:2700–2713. https://doi.org/10.1074/jbc.RA118.007095

Romagnoli G, Nisini R, Chiani P, Mariotti S, Teloni R, Cassone A, Torosantucci A (2004) The interaction of human dendritic cells with yeast and germ-tube forms of. J Leukoc Biol 75:117–126. https://doi.org/10.1189/jlb.0503226.1

Rosario-colon J, Eberle K, Adams A, Courville E, Xin H (2021) Candida cell-surface-specific monoclonal antibodies protect mice against candida auris invasive infection. Int J Mol Sci 22:6162. https://doi.org/10.3390/ijms22116162

Rudkin FM, Raziunaite I, Workman H, Essono S, Belmonte R, MacCallum DM, Johnson EM, Silva LM, Palma AS, Feizi T et al (2018) Single human B cell-derived monoclonal anti-Candida antibodies enhance phagocytosis and protect against disseminated candidiasis. Nat Commun 9:5288. https://doi.org/10.1038/s41467-018-07738-1

Sandini S, La Valle R, Deaglio S, Malavasi F, Cassone A, De Bernardis F (2011) A highly immunogenic recombinant and truncated protein of the secreted aspartic proteases family (rSap2t) of Candida albicans as a mucosal anticandidal vaccine. FEMS Immunol Med Microbiol 62:215–224. https://doi.org/10.1111/j.1574-695X.2011.00802.x

Sanglard D, Hube B, Monod M, Odds FC, Gow NAR (1997) A triple deletion of the secreted aspartyl proteinase genes SAP4, SAP5, and SAP6 of Candida albicans causes attenuated virulence. Infect Immun 65:3539–3546. https://doi.org/10.1128/iai.65.9.3539-3546.1997

Santoni G, Boccanera M, Adriani D, Lucciarini R, Amantini C, Morrone S, Cassone A, De Bernardis F (2002) Immune cell-mediated protection against vaginal candidiasis: evidence for a major role of vaginal CD4+ T cells and possible participation of other local lymphocyte effectors. Infect Immun 70:4791–4797. https://doi.org/10.1128/IAI.70.9.4791-4797.2002

Satala D, Karkowska-Kuleta J, Zelazna A, Rapala-Kozik M, Kozik A (2020) Moonlighting proteins at the candidal cell surface. Microorganisms 8:1–25. https://doi.org/10.3390/microorganisms8071046

Schatzman SS, Peterson RL, Teka M, He B, Cabelli DE, Cormack BP, Culotta VC (2020) Copper-only superoxide dismutase enzymes and iron starvation stress in Candida fungal pathogens. J Biol Chem 295:570–583. https://doi.org/10.1074/jbc.RA119.011084

Schmidt CS, White CJ, Ibrahim AS, Filler SG, Fu Y, Yeaman MR, Edwards JE, Hennessey JP (2012) NDV-3, a recombinant alum-adjuvanted vaccine for Candida and Staphylococcus aureus, is safe and immunogenic in healthy adults. Vaccine 30:7594–7600. https://doi.org/10.1016/j.vaccine.2012.10.038

Selvakumar D, Karim N, Miyamoto M, Furuichi Y, Komiyama T (2006) Recombinant single-chain anti-idiotypic antibody: An effective fungal β-1, 3-glucan synthase inhibitor. Biological and Pharmaceutical Bulletin 29(9):1848–1853

Sendid B, Lecointe K, Collot M, Danzé PM, Damiens S, Drucbert AS, Fradin C, Vilcot JP, Grenouillet F, Dubar F et al (2021) Dissection of the anti-Candida albicans mannan immune response using synthetic oligomannosides reveals unique properties of β-1,2 mannotriose protective epitopes. Sci Rep 11:1–14. https://doi.org/10.1038/s41598-021-90402-4

Shi H, Dong S, Zhang X, Chen X, Gao X, Wang L (2018) Phage vaccines displaying YGKDVKDLFDYAQE epitope induce protection against systemic candidiasis in mouse model. Vaccine 36:5717–5724. https://doi.org/10.1016/j.vaccine.2018.08.011

Shibasaki S, Aoki W, Nomura T, Miyoshi A, Tafuku S, Sewaki T, Ueda M (2013) An oral vaccine against candidiasis generated by a yeast molecular display system. Pathog Dis 69:262–268

Shibasaki S, Aoki W, Nomura T, Karasaki M, Sewaki T, Ueda M (2014) Evaluation of Mdh1 protein as an antigenic candidate for a vaccine against candidiasis. Biocontrol Sci 19:51–55. https://doi.org/10.4265/bio.19.51

Shibasaki S, Karasaki M, Aoki W, Ueda M (2018) Molecular and physiological study of Candida albicans by quantitative proteome analysis. Proteomes 6:34. https://doi.org/10.3390/proteomes6030034

Shukla M, Rohatgi S (2020) Vaccination with secreted aspartyl proteinase 2 protein from candida parapsilosis can enhance survival of mice during c. tropicalis-mediated systemic candidiasis. Infect Immun 88:e00312–e00320. https://doi.org/10.1128/IAI.00312-20

Shukla M, Chandley P, Rohatgi S (2021) The role of B-cells and antibodies against *Candida* vaccine antigens in invasive candidiasis. Vaccine 9:1159. https://doi.org/10.3390/vaccines9101159

Silva RC, Padovan ACB, Pimenta DC, Ferreira RC, da Silva CV, Briones MRS (2014) Extracellular enolase of Candida albicans is involved in colonization of mammalian intestinal epithelium. Front Cell Infect Microbiol 4:1–9. https://doi.org/10.3389/fcimb.2014.00066

Singh S, Uppuluri P, Alqarihi A, Elhassan H, French S, Lockhart SR, Chiller T, Edwards JE, Ibrahim AS (2019) The NDV-3A vaccine protects mice from multidrug resistant Candida auris infection. PLoS Pathog 15:e1007460. https://doi.org/10.1101/465096

Sinha BK, Prasad S, Monga DP (1987) Studies of the role of B-cells in the resistance of mice to experimental candidiasis. Zentralbl Bakteriol Mikrobiol Hyg A 266:316–322. https://doi.org/10.1016/S0176-6724(87)80045-7

Smolenski G, Sullivan PA, Cutfield SM, Cutfield JF (1996) Analysis of secreted aspartic proteinases from Candida albicans : purification and characterization of individual Sapl , Sap2 and Sap3 isoenzymes. Microbiology 143:349–356

Spellberg B, Ibrahim AS, Edwards JE, Filler SG (2005) Mice with disseminated candidiasis die of progressive sepsis. J Infect Dis 192:336–343. https://doi.org/10.1086/430952

Spellberg B, Ibrahim AS, Lin L, Avanesian V, Fu Y, Lipke P, Otoo H, Ho T, Edwards JE (2008) Antibody titer threshold predicts anti-candidal vaccine efficacy even though the mechanism of protection is induction of cell-mediated immunity. J Infect Dis 197:967–971. https://doi.org/10.1086/529204

Staib P, Lermann U, Blaß-Warmuth J, Degel B, Würzner R, Monod M, Schirmeister T, Morschhäuser J (2008) Tetracycline-inducible expression of individual secreted aspartic proteases in Candida albicans allows isoenzyme-specific inhibitor screening. Antimicrob Agents Chemother 52:146–156. https://doi.org/10.1128/AAC.01072-07

Sundstrom P, Jensen J, Balish E (1994) Humoral and cellular immune responses to enolase after alimentary tract colonization or intravenous immunization with candida albicans. J Infect Dis 170:390–395. https://doi.org/10.1093/infdis/170.2.390

Swoboda RK, Bertram G, Hollander H, Greenspan D, Greenspan JS, Gow NAR, Gooday GW, Brown AJP (1993) Glycolytic enzymes of Candida albicans are nonubiquitous immunogens during candidiasis. Infect Immun 61:4263–4271. https://doi.org/10.1128/iai.61.10.4263-4271.1993

Theiss S, Ishdorj G, Brenot A, Kretschmar M, Lan CY, Nichterlein T, Hacker J, Nigam S, Agabian N, Köhler GA (2006) Inactivation of the phospholipase B gene PLB5 in wild-type Candida albicans reduces cell-associated phospholipase A2 activity and attenuates virulence. Int J Med Microbiol 296:405–420. https://doi.org/10.1038/jid.2014.371

Tian R, Fu M, Zhang Z, Ren J, An J, Liu Y, Li W (2013) In Situ IgM production and clonal expansion of B-1 cells in peritoneal cavity promote elimination of C. albicans infection in IgH transgenic mice with VH derived from a natural antibody. PLoS One 8:e60779. https://doi.org/10.1371/journal.pone.0060779

Torosantucci A, Bromuro C, Chiani P, De Bernardis F, Berti F, Galli C, Norelli F, Bellucci C, Polonelli L, Costantino P et al (2005) A novel glyco-conjugate vaccine against fungal pathogens. J Exp Med 202:597–606. https://doi.org/10.1084/jem.20050749

Torosantucci A, Chiani P, Bromuro C, De Bernardis F, Palma AS, Liu Y, Minogna G, Maras B, Colone M, Stringaro A et al (2009) Protection by anti-β-glucan antibodies is associated with

restricted β-1,3 glucan binding specificity and inhibition of fungal growth and adherence. PLoS One 4:e5392. https://doi.org/10.1371/journal.pone.0005392
Torosantucci A, Tumbarello M, Bromuro C, Chiani P, Posteraro B, Sanguinetti M, Cauda R, Cassone A (2017) Antibodies against a β-glucan-protein complex of Candida albicans and its potential as indicator of protective immunity in candidemic patients. Sci Rep 7:1–9. https://doi.org/10.1038/s41598-017-02977-6
Tsoni SV, Kerrigan AM, Marakalala MJ, Srinivasan N, Duffield M, Taylor PR, Botto M, Steele C, Brown GD (2009) Complement C3 plays an essential role in the control of opportunistic fungal infections. Infect Immun 77:3679–3685. https://doi.org/10.1128/IAI.00233-09
Tsuchimori N, Sharkey LL, Fonzi WA, French SW, Edwards JE, Filler SG (2000) Reduced virulence of HWP1-deficient mutants of Candida albicans and their interactions with host cells. Infect Immun 68:1997–2002. https://doi.org/10.1128/IAI.68.4.1997-2002.2000
Uppuluri P, Lin L, Alqarihi A, Luo G, Youssef EG, Alkhazraji S, Yount NY, Ibrahim BA, Bolaris MA, Edwards JE et al (2018a) The Hyrl protein from the fungus Candida albicans is a cross kingdom immunotherapeutic target for Acinetobacter bacterial infection. PLoS Pathog 14:1–25. https://doi.org/10.1371/journal.ppat.1007056
Uppuluri P, Singh S, Alqarihi A, Schmidt CS, Hennessey JP, Yeaman MR, Filler SG, Edwards JE, Ibrahim AS (2018b) Human anti-Als3p antibodies are surrogate markers of NDV-3A vaccine efficacy against recurrent vulvovaginal candidiasis. Front Immunol 9:1349. https://doi.org/10.3389/fimmu.2018.01349
Van De Veerdonk FL, Lauwerys B, Marijnissen RJ, Timmermans K, Di Padova F, Koenders MI, Gutierrez-Roelens I, Durez P, Netea MG, Van Der Meer JWM et al (2011) The anti-CD20 antibody rituximab reduces the Th17 cell response. Arthritis Rheum 63:1507–1516. https://doi.org/10.1002/art.30314
Van de Veerdonk FL, Kullberg BJ, Netea MG (2012) Adjunctive immunotherapy with recombinant cytokines for the treatment of disseminated candidiasis. Clin Microbiol Infect 18(2):112–119
Van Deventer HJ, Goessens WH, Van Vliet AJ, Verbrugh HA (1996) Anti-enolase antibodies partially protective against systemic candidiasis in mice. Clin Microbiol Infect 2:36–43. https://doi.org/10.1111/j.1469-0691.1996.tb00198.x
Vilanova M, Teixeira L, Caramalho Í, Torrado E, Marques A, Madureira P, Ribeiro A, Ferreira P, Gama M, Demengeot J (2004) Protection against systemic candidiasis in mice immunized with secreted aspartic proteinase 2. Immunology 111:334–342. https://doi.org/10.1111/j.1365-2567.2004.01819.x
Voigt J, Malone DFG, Dias J, Leeansyah E, Björkström NK, Ljunggren HG, Gröbe L, Klawonn F, Heyner M, Sandberg JK et al (2018) Proteome analysis of human CD56 neg NK cells reveals a homogeneous phenotype surprisingly similar to CD56 dim NK cells. Eur J Immunol 48:1456–1469. https://doi.org/10.1002/eji.201747450
Vonk AG, Netea MG, Van Der Meer JWM, Kullberg BJ (2006) Host defence against disseminated Candida albicans infection and implications for antifungal immunotherapy. Expert Opin Biol Ther 6:891–903. https://doi.org/10.1517/14712598.6.9.891
Vudhichamnong K, Walker DM, Ryley HC (1982) The effect of secretory immunoglobulin A on the in-vitro adherence of the yeast Candida albicans to human oral epithelial cells. Arch Oral Biol 27:617–621
Wagner RD, Vazquez-Torres A, Jones-Carson J, Warner T, Balish E (1996) B cell knockout mice are resistant to mucosal and systemic candidiasis of endogenous origin but susceptible to experimental systemic candidiasis. J Infect Dis 174:589–597. https://doi.org/10.1093/infdis/174.3.589
Wan Tso GH, Reales-Calderon JA, Pavelka N (2018) The elusive anti-Candida vaccine: lessons from the past and opportunities for the future. Front Immunol 9:897. https://doi.org/10.3389/fimmu.2018.00897
Wang Y, Su Q, Dong S, Shi H, Gao X, Wang L (2014) Hybrid phage displaying SLAQVKYTSASSI induces protection against Candida albicans challenge in BALB/c mice. Hum Vaccin Immunother 10:1057–1063. https://doi.org/10.4161/hv.27714

Wang XJ, Sui X, Yan L, Wang Y, Cao YB, Jiang YY (2015) Vaccines in the treatment of invasive candidiasis. Virulence 6:309–315. https://doi.org/10.4161/21505594.2014.983015

Warnock DW (2007) Trends in the epidemiology of invasive fungal infections. Japanese J Med Mycol 48:1–12. https://doi.org/10.3314/jjmm.48.1

Whibley N, Gaffen SL (2015) Beyond Candida albicans: mechanisms of immunity to non-albicans Candida species. Cytokine 76:42–52. https://doi.org/10.1016/j.cyto.2015.07.025

White TC, Agabian N (1995) Candida albicans secreted aspartyl proteinases: isoenzyme pattern is determined by cell type, and levels are determined by environmental factors. J Bacteriol 177:5215–5221

White TC, Miyasaki SH, Agabian N (1993) Three distinct secreted aspartyl proteinases in Candida albicans. J Bacteriol 175:6126–6133. https://doi.org/10.1128/jb.175.19.6126-6133.1993

Wu X, Lipinski T, Carrel FR, Bailey JJ, Bundle DR (2007) Synthesis and immunochemical studies on a Candida albicans cluster glycoconjugate vaccine. Org Biomol Chem 5:3477–3485. https://doi.org/10.1039/b709912f

Wu X, Lipinski T, Paszkiewicz E, Bundle DR (2008) Synthesis and immunochemical characterization of 5-linked glycoconjugate vaccines against Candida albicans. Chem Eur J 14:6474–6482. https://doi.org/10.1002/chem.200800352

Xin H (2016) Active immunizations with peptide-DC vaccines and passive transfer with antibodies protect neutropenic mice against disseminated candidiasis. Vaccine 34:245–251. https://doi.org/10.1016/j.vaccine.2015.11.035

Xin H, Cutler JE (2011) Vaccine and monoclonal antibody that enhance mouse resistance to candidiasis. Clin Vaccine Immunol 18:1656–1667. https://doi.org/10.1128/CVI.05215-11

Xin H, Dziadek S, Bundle DR, Cutler JE (2008) Synthetic glycopeptide vaccines combining β-mannan and peptide epitopes induce protection against candidiasis. Proc Natl Acad Sci USA 105:13526–13531. https://doi.org/10.1073/pnas.0803195105

Xin H, Cartmell J, Bailey JJ, Dziadek S, Bundle DR, Cutler JE (2012) Self-adjuvanting glycopeptide conjugate vaccine against disseminated candidiasis. PLoS One 7:e35106. https://doi.org/10.1371/journal.pone.0035106

Xin H, Glee P, Adams A, Mohiuddin F, Eberle K (2019) Design of a mimotope-peptide based double epitope vaccine against disseminated candidiasis. Vaccine 37:2430–2438. https://doi.org/10.1016/j.vaccine.2019.03.061

Yeaman MR, Filler SG, Chaili S, Barr K, Wang H, Kupferwasser D, Hennessey JP, Fu Y, Schmidt CS, Edwards JE et al (2014) Mechanisms of NDV-3 vaccine efficacy in MRSA skin versus invasive infection. Proc Natl Acad Sci USA 111:E5555–E5563. https://doi.org/10.1073/pnas.1415610111

Youssef EG, Zhang L, Alkhazraji S, Gebremariam T, Singh S, Yount NY, Yeaman MR, Uppuluri P, Ibrahim AS (2020) Monoclonal IgM antibodies targeting Candida albicans Hyr1 provide cross-kingdom protection against gram-negative bacteria. Front Immunol 11:6–8. https://doi.org/10.3389/fimmu.2020.00076

Zhang MX, Kozel TR (1997) Mannan-specific immunoglobulin G antibodies in normal human serum mediate classical pathway initiation of C3 binding to Candida albicans. Infect Immun 65:3822–3827. https://doi.org/10.1128/iai.66.10.4845-4850.1998

Zhang MX, Cutler JE, Han Y, Kozel TR (1998) Contrasting roles of mannan-specific monoclonal immunoglobulin M antibodies in the activation of classical and alternative pathways by Candida albicans. Infect Immun 66:6027–6029. https://doi.org/10.1128/iai.66.12.6027-6029.1998

Zhang MX, Bohlman MC, Itatani C, Burton DR, Parren PWHI, St. Jeor SC, Kozel TR (2006) Human recombinant antimannan immunoglobulin G1 antibody confers resistance to hematogenously disseminated candidiasis in mice. Infect Immun 74:362–369. https://doi.org/10.1128/IAI.74.1.362-369.2006

Zito A, Bromuro C, Mandili G, Chiani P, Horenstein AL, Malavasi F, Cauda R, Cassone A, Torosantucci A (2016) A murine, bispecific monoclonal antibody simultaneously recognizing β-glucan and MP65 determinants in Candida species. PLoS One 11:1–15. https://doi.org/10.1371/journal.pone.0148714

# Index




A. Gomes Rodrigues, A. Gupta (eds.), *Fungal Infections*, Fungal Biology,
https://doi.org/10.1007/978-3-032-06014-3

MIX
Papier aus verantwortungsvollen Quellen
Paper from responsible sources
FSC® C105338

If you have any concerns about our products,
you can contact us on
**ProductSafety@springernature.com**

In case Publisher is established outside the EU,
the EU authorized representative is:
**Springer Nature Customer Service Center GmbH**
**Europaplatz 3, 69115 Heidelberg, Germany**

Printed by Libri Plureos GmbH
in Hamburg, Germany